CAMBRIDGE LIBRARY COLLECTION

Books of enduring scholarly value

Botany and Horticulture

Until the nineteenth century, the investigation of natural phenomena, plants and animals was considered either the preserve of elite scholars or a pastime for the leisured upper classes. As increasing academic rigour and systematisation was brought to the study of 'natural history', its subdisciplines were adopted into university curricula, and learned societies (such as the Royal Horticultural Society, founded in 1804) were established to support research in these areas. A related development was strong enthusiasm for exotic garden plants, which resulted in plant collecting expeditions to every corner of the globe, sometimes with tragic consequences. This series includes accounts of some of those expeditions, detailed reference works on the flora of different regions, and practical advice for amateur and professional gardeners.

Experiments and Observations
Concerning Agriculture and the Weather

William Marshall (1745–1818), from farming stock, became a farmer and then estate manager and land agent after several years conducting business in the West Indies. This 1779 book (one of his earliest) describes his observations and experiments on his farm in Surrey (which he later had to give up because of his partner's bankruptcy). A description of the size, soil type and aspect of his various fields is followed by a summary of the experiments he carried out – mostly simple ones, such as comparing results if seeded fields were rolled or not. Diary records over two years for each crop are given, with areas sown, soil conditions and weather data. A chapter is devoted to weather prognostications, and another to day-to-day farm management and accounts. The systematic reporting of his findings will, Marshall hopes, be of use to others, and provides interesting insights into the beginnings of scientifically based agriculture.

Experiments and Observations
Concerning
Agriculture and the Weather

WILLIAM MARSHALL

CAMBRIDGE
UNIVERSITY PRESS

CAMBRIDGE
UNIVERSITY PRESS

University Printing House, Cambridge, CB2 8BS, United Kingdom

Cambridge University Press is part of the University of Cambridge.
It furthers the University's mission by disseminating knowledge in the pursuit of
education, learning and research at the highest international levels of excellence.

www.cambridge.org
Information on this title: www.cambridge.org/9781108075831

© in this compilation Cambridge University Press 2015

This edition first published 1779
This digitally printed version 2015

ISBN 978-1-108-07583-1 Paperback

This book reproduces the text of the original edition. The content and language reflect
the beliefs, practices and terminology of their time, and have not been updated.

Cambridge University Press wishes to make clear that the book, unless originally published
by Cambridge, is not being republished by, in association or collaboration with,
or with the endorsement or approval of, the original publisher or its successors in title.

EXPERIMENTS

AND

OBSERVATIONS

CONCERNING

AGRICULTURE

AND THE

WEATHER.

By Mr. MARSHALL,

AUTHOR OF THE

MINUTES OF AGRICULTURE.

LONDON:

Printed for J. DODSLEY, in Pall-Mall.

M.DCC.LXXIX.

ADVERTISEMENT.

WHEN, in May laſt, the Writer of this Advertiſement ſubmitted his MINUTES OF AGRICULTURE to the Public, he had not the moſt diſtant intentions of again communicating his ſentiments, publicly, in ſo ſhort a time. A variety of thoſe untoward incidents which Fate thinks fit to conceal from human foreſight has, however, tended, or, as it were, conſpired, to counteract what he had thought laudable intentions.

The Farms, the ſubject of the above-mentioned Publication, as well as of the ſucceeding ſheets, were juſt arriving at the ſtate to which he had been four years ſtruggling to bring them. The hurry which neceſſarily attends the outſet of Farming began to ſubſide, and the GENERAL MANAGEMENT began to flow in regular channels: he had therefore more leiſure and inclination to attend to the accuracies of the MINUTIAL MANAGEMENT; and to the Making, Regiſtering, and Obſerving the reſults of COMPARATIVE EXPERIMENTS. Added to this, he had fallen upon a *new*, and what he eſteemed an *improved* Mode of *Regiſtering* his EXPERIMENTS; and likewiſe on a more *ſcientific* Mode of *Recording* his OBSERVATIONS; and at the expiration of the principal Leaſe, which, in the ordinary courſe of things, would have happened in 1782, he intended to have reviſed his Acquiſitions, and to have publiſhed ſuch as might then have been deemed worthy of publication.

His intentions, however, have been fruſtrated; and inſtead of *ſix*, he is unable to offer more than *two years' experience:* Nevertheleſs, *as the Sketch of a Plan of acquiring Agricultural Information from Self-practice*, the intended Work would not, probably, have been more explicit than that which is now offered to the Public.

Here the Writer would have cloſed his Apology, had he not been prompted by a deſire of turning aſide thoſe unfavourable ſurmiſes, which malevolence and miſinformation are ever inclined to propagate; eſpecially when men's actions are placed in thoſe perilous predicaments, which the moſt cautious circumſpection and the beſt intentions are ſometimes unable to avoid. The Writer is the more anxious to do himſelf this juſtice here, as this is, perhaps, the only opportunity he may have of doing it with *propriety*; it is, indeed, in ſome meaſure *neceſſary*, that every AGRICULTURAL WRITER ſhould faithfully communicate his MEMOIRS AS A FARMER.

He has already ſaid, in another place, that he was *born a Farmer* : he might have added, that he can trace his blood through the veins of AGRICULTURISTS, for upwards of four hundred years. He has already ſaid, too, that he was *bred to traffic :* he did, it is true, *wander* in the ways of Commerce fourteen years; and until the year 1771, he retained, in ſome degree, the enthuſiaſtic idea with which he ſet out, of—*making a Fortune*.

In that year, however, his train of thinking was not leſs altered than his conſtitution, by a violent fit of illneſs. His recovery was ſo ſingular, that he has ever ſince conſidered it as a *ſecond birth*; and his *ſecond life* as a ſpecial gift of Providence. Theſe circumſtances are mentioned, as probably in them originate the preſent Memoirs.

Prior

ADVERTISEMENT.

Prior to this time, however, he had wifhed earneftly to return to a rural life; but the inconfiftency of quitting, with feeming wantonnefs, the path he had chofen, and other more effential reafons, iu a great meafure ftifled the wifh; but being now fet free from the charms of *Fortune-Hunting*, by being convinced of the *uncertainty* of human life; and being foon after enabled, by the demife of his Father, to profecute his intentions with greater propriety, he no longer hefitated to look out for a Farm.

He had, indeed, many other inducements, befides thofe mentioned, to purfue the plan he had drawn: his whole life had been, more-or-lefs, a life of *Study*; he had rambled, in a defultory way, over the whole field of Science; his paffion for *fcientific knowledge* had long been fo prevailing, that nothing but the ambition of becoming *rich* could have ftemmed it. This bubble, however, being now broke, his defire for Study confequently encreafed, and the *Exchange*, the *Coffee-Houfe*, and the *Defk*, he found to be wholly inimical to thofe abftrufe inveftigations which had ever been his favourite amufements; and which, added to the bufinefs and the air of London, had probably laid the foundation of his fevere illnefs. *The active life of a Hufbandman* feemed, of all others, moft eligible to counteract the evil tendency of fedentary contemplation; and a *fmall Farm* appeared the beft adapted to furnifh corporeal exercife, without requiring much of the mind's affiftance.

There yet remain other reafons for his quitting the *bufinefs of Infurance* for the *avocation of Agriculture*; and as he has ever made it a practice to reduce his intentions to writing before he executes them, he might here have tranfcribed a Minute he made on the fubject of his leaving the Metropolis; but left he fhould give offence to many whom he would not voluntarily offend, he will only fay, that there are certain fenfations of the human mind, which fome *are*, and others *are not* fufceptible of; and which fome *can*, and others *cannot* facrifice: added to this, there is a fpecies of *Independency* to be found in Agriculture, which in Trade *muft not*, which *cannot*, on the leading principles of Commerce, be looked for.

The Profits of Farming *he knew* to be *fmall*; but he had refolved that Industry even to *Labour*, and Economy down to the *fimpleft frugality*, fhould make up the deficiencies in emolument. Befides, the neighbourhood of London was the place he had pitched upon for a Farm; and his intentions were to have retained his bufinefs in London (which was principally tranfacted from twelve to three o'clock), until he had found himfelf *feated* in his Farm.

While he was looking out for a Farm on thofe terms, and with thefe intentions, he had an offer made him by an intimate Friend, who had then in his poffeffion a confiderable quantity of land, to become his *Partner in Farming*.

It fometimes happens, that thofe things which we affect moft to defpife, we the fooneft embrace. The Writer had formed the moft unfavourable idea of *Copartnerfhips*. He had, indeed, had various opportunities of feeing their evil confequences; and had formed a refolution never to enter into one with any man, let the profpect be ever fo lucrative. The profpect, however, which he had now before him was not Lucre: Health, Science, and the Converse of Nature, were the objects which prefented themfelves. Befides, he had looked out in vain for the Farm to his wifh; the false Spirit of Farming then fhone out with meridian fplendour, and all ranks and defcriptions of people had fled to the country, to enjoy the *genial funfhine*; but had not yet difcovered, that they had fubjected themfelves to the influence of a *baleful meteor*.

In

ADVERTISEMENT.

In fhort, Health pleaded for the Country, and Friendfhip for the connexion ; and who, in the year 1774, would not have formed a reputable connexion with Mr. ***** *****? Who was more beloved as a Friend? Who more refpected as a Merchant?

The connexion was formed : the *Leafes* were Mr. *****'s, and the *entire Management* became the department of the Writer.

The Farms being fcattered, and a principal part of them being very much out of condition, he found himfelf engaged in a more *ferious tafk* than he had been aware of : he faw full employment for *all* the attendance and attention he was poffeffed of ; and he prefently found himfelf obliged to give up not only his bufinefs in London, but, what was a more *fenfible* lofs, that purfuit of *General Knowledge* which had made a principal part of his defire for a country life. He had intended to have committed the *Minutial Management*, at leaft, to the care of a *Bailiff*, and to have attended *himfelf* to the *General Management* alone : but he prefently difcovered his error ; he found that the MINUTIAL MANAGEMENT conftitutes the *vitals* of Agriculture, and is, of all others, the leaft fit to be left, folely, to the care of a fervant ; he therefore gave up his whole felf to Farming.

His want of experience he was refolved to fupply by unremitted application : he not only made Agriculture his whole *bufinefs*, but his fole *amufement* and his only *ftudy*. In order that he might the more fpeedily become, what of all things he now wifhed to be, *a good Farmer*, he clofely attended to the *mereft Minutiæ*, and endeavoured to make himfelf mafter of *every manual operation* ; and that he might not lofe any part of the experience he gained, he carefully committed his fucceffes and mifcarriages to writing.

By the intenfe application neceffary to the profecution of this plan, he foon gained an infight, not only into the *Management* of a Farm, but into the *Bufinefs* of Farming. The expences of a Farm, during the firft two or three years, are, no doubt, much fuperior to thofe which are afterwards requifite ; but before the expiration of the fecond year, he was neverthelefs enabled, by the *Minute Accounts* which he had kept, to calculate with fome degree of certainty, that the *Profits* of the Farms he was poffeffed of, could not, in their moft improved ftate, be confiderable. Indeed, it appeared to him a doubt, whether they would ever be able to difcharge the heavy Rent and Taxes with which they were faddled. He had perceived fome of the foils to be bad, fome only indifferent, and the whole fo detached and fcattered, that it was difficult to apply them to a profitable purpofe.

In 1776, Mr. ***** went to refide in a diftant part of the Ifland, and confequently wifhed to break off every connexion with Surry. This, however, was not eafily to be done : the Rents of the Farms were too high, and the unexpired term of the principal Leafe was fhort. Added to this, the *rage* for Farming which had glowed, or rather burnt out fo furioufly of late, had now begun to abate ; and the Writer knew too well the difadvantages above enumerated, to take the leafes upon himfelf : therefore, *from this period*, he confidered himfelf as one whofe beft offices were due to *Friendfhip* ; and he reflects with pleafure, that this was a frefh ftimulus to application. An additional incentive, however, was not wanted ; for he was now beginning to fee the expected fruits of his labour. He was beginning to fee his plans fucceed ; his

crops

ADVERTISEMENT.

crops increafe; his oxen become tractable ; and his Implements improve. In fhort, his Farm was become his *place of amufement*, and his Minute-Book his *Companion*: Retirement of courfe became more and more agreeable.

The paffion for *General Knowledge* was now wholly fuperfeded by the *Science of Agri-culture*. To be an *Author*, indeed, had been long his ambition ; but to be an *Agri-cultural Author* was, he declares, an idea which never occurred to him, until he had been near three years a Farmer ; when, being convinced of the difparity between *actual Farming*, and that *written Agriculture* which he had formerly read, he wifhed to convey to *Agricultural Readers* a *real fketch of private Agriculture* ; and could not hit upon any other mode of doing this, than by publifhing his Minutes literally as he made them.

In Autumn 1777, he therefore began to prepare his Minutes for the Prefs ; but having at the fame time the bufinefs of the Farm to attend to, and the Digest prov-ing more copious than was at firft intended, it was the beginning of March 1778, before the Copy was finifhed.

He was now arriving at what he efteemed a defirable period : the difficulties of Farm-ing were paft, and the drudgery of publication drawing near to a clofe ; with full four years of happinefs in ftore. What added confiderably to his profpect, the plan of ma-nagement he had been cautioufly drawing *from his own experience*, was arrived at fuch a ftate of forwardnefs, as no longer to be in danger of being marred by *adventitious intercourfe*.

But how fhort is human forefight ! This bright appearance of approaching happi-nefs was one of thofe *ferene mornings* which precede, as it were to heighten the ef-fects of, approaching Earthquakes ! for the very day on which the Writer put a finifhing hand to the *Copy* of the Digest, the fame day Mr. ***** became fub-jected to a certain *commercial Law*, which diffolves, conditionally, all human agree-ments ; and the Farms being much too high-rented, they were in prudence given up to their refpective Landlords.

Nor did the Writer's difficulties end here. He wifhed to have retained in his poffeffion a portion of thofe Farms which had during four years engaged every hour of his attention. He was the more anxious, as this was the part which had been peculiarly his *home*, and had of courfe engaged a more particular fhare of his attendance. If he wanted ex-ercife or amufement, he found it here. There is fcarcely a field which he has not plow-ed, nor a hedge he has not more-or-lefs trimmed with his own hands. It confifted of four Arable Divifions *, with a proportionate quantity of Pafture and *Garden-Ground* † ; and is of fuch a fize, that with four oxen he could have cultivated it (with-out fuffering a horfe to ftep upon it) in a manner to his wifh. The Farm-Yard, too, was a *child of his own*. In fhort, befides its having become his *home*, it was (the wet-nefs of its fituation apart) the kind of Farm he had ever wifhed for. He could picture to himfelf the days he fhould henceforward pafs in manual labour, and the evenings he fhould enjoy in fcientific refearches.

A variety of obftacles, however, prefented themfelves. This Farm having been rented at nearly twice its value, it was reafonable to fuppofe that the Proprietor would not let it down to its real worth ; and the Writer, who is convinced himfelf, and has endeavoured to convince others, that every man who advances, or aids in advancing the rents of lands *above* their *value*, is a Enemy to his Country, *could not* give

* F. G. H. I.   † S. and T.

give

give more for it than what he knew, from five years experience, to be its full value: Befides, the Proprietor was a man of *fuch* principles, and had behaved in *fuch* a manner, as to render it impoffible for the Writer to enter into any connexion with *him*.

This obftacle, however, was timely removed: for the Eftate fell, *as Eftates are wont to do*, into the hands of *Bankers*. But here a frefh barrier arofe: The Eftate came into *their* hands *in the way of Trade*; and, as *fair Tradefmen*, they had a right to make the moft of it. In fhort, the Farm was eftimated at one-third more than what the Writer *knew*, and ftill *knows*, to be its *full value*; and lett to a *Neighbour*, ——whofe *birth, parentage*, and *education*; *life, character*, and ——BEHAVIOUR, are, jointly and feverally—not worth recording.

No apology is made for this detail, (which to the generality of Readers may appear tedious and uninterefting) except, that a man who has few of the good things of this world but the good opinion of others to be anxious for, has a fpecial privilege of being even *tedioufly* tenacious of that opinion. And, indeed, the Author thinks it in fome meafure neceffary, that his Readers in general fhould be fatisfied that he neither took up, nor laid down, FARMING, *capricioufly :* and as a further proof of his confiftency he acquaints them, that although he is not at prefent a *Farmer*, he has not loft fight of SCIENTIFIC AGRICULTURE.

THE

INTRODUCTION

TO THE

EXPERIMENTS.

IF we compare Human Knowledge with OMNISCIENCE, the former is narrowly circumfcribed. *Theory, Hypothefis,* and *Opinion,* it is true, abound; but the CERTAIN KNOWLEDGE *actually poffeffed* even by the moft enlightened underftanding, might be compreffed into a narrow compafs.

It muft be acknowledged, however, (and it ought to be acknowledged with fingular gratitude by the prefent inhabitants of civilized nations) that it is matter of aftonifhment how a Being fo *limited* and fo *tranfient* as Man is, can make the advances which at various periods he has done towards UNIVERSAL KNOWLEDGE. The degree of elevation at which ASTRONOMY, COSMOGRAPHY, BOTANY, THE MATHEMATICS, MECHANICS, ARCHITECTURE, NAVIGATION, THE FINE ARTS, and LITERATURE, now foar, would, in a lefs enlightened age, have been confidered as a height unattainable by the human underftanding.

But amidft all the revolutions of human ARTS and SCIENCES, is it not worthy of remark that AGRICULTURE fhould never have been treated as a SCIENCE? that it fhould never have been profeffedly confidered as a branch of EXPERIMENTAL PHILOSOPHY? Yet, inexplicable as this may be, it would, perhaps, be difficult, from the numberlefs Volumes which are extant on that fubject, to extract one *authentic* fheet of COMPARATIVE EXPERIMENTS. This is the more to be wondered at, as almoft every GREAT NATION and many GREAT MEN have paid a particular attention to AGRICULTURE. Even VIRGIL, who probably wrote under the mediate patronage of a GREAT PRINCE, and who was not only a *practical* agriculturift, but may be fuppofed to have ftudied the *written* Agriculture extant in his time, neither appears to have made himfelf, nor recommends to others, A COURSE OF SCIENTIFIC EXPERIMENTS.

The

THE INTRODUCTION

THE UTILITY OF EXPERIMENTING, however, muft be obvious to every one; for although a fortuitous INCIDENT may be equally inftructive as the looked-for refult of an EXPERIMENT; yet the former being fubject to *chance*, may be few; while the latter can be encreafed at the *will* of the Experimentalift. EXPERIMENTING, therefore, muft be peculiarly ufeful to the NOVITIAL AGRICULTURIST; and, indeed, to every Farmer, let him be ever fo fkilful in his profeffion, at his firft entrance into a NEW FARM.

THE MODE OF MAKING EXPERIMENTS—requires little explanation. It may be proper to obferve, however, that CIRCUMSPECTION and ACCURACY are indifpenfibly neceffary to the operation; the former to guard againft any *diffimilarity* of *Soil*, *Seed*, &c &c. with which the Experiment is about to be made; and the latter to mark minutely the *fcene of Experiment*. When this lies in the *Field*, LABELLED STUMPS are very *convenient*: but they fhould not be *implicitly depended upon*, being liable to be *removed*, either by accident or intention; the *place* fhould therefore be identified likewife by the number of Lands, quantity of Rods, &c. In fhort, TO MAKE AN AUTHENTIC EXPERIMENT, an *identity* of PLACE, TIME, ELEMENT, and PROCESS, muft be ftrictly obferved in every particular, excepting only the INTENDED DIFFERENCE which conftitutes the Experiment. Nor can the Experiment be *authentic*, if the Procefs be in any inftance left to an *Agent:* it muft be performed by the immediate hand, or under the immediate eye of the EXPERIMENTALIST.

THE MODE OF REGISTERING EXPERIMENTS—is of fome importance. The *Accuracy of Making* is loft, if the Experiment be not *fully and accurately regiftered*. The method which the Writer adopted two years ago, and which he has not yet found any reafon to alter, may be feen in the following Regifter. The PROCESSES he regifters chronologically as they are made; leaving blank receffes on the oppofite fide of the fame page, in order to receive the refpective RESULTS. And in order to render each Experiment ftill more perfpicuous, he numbers it; refers to the Field, &c. in which it is made; gives it a *head*, and briefly minutes his INTENTION for making it. This he does by way of *interrogative*, in order that he may have an opportu-

tunity

tunity of giving a *concife, anfwer*, comprehending the *amount* of the Refult. p The PROCESS and RESULT would feem barren and abftrufe, were they not amplified and elucidated by the INTENTION.

At firft fight, perhaps, there may appear an aukwardnefs in THE ARRANGEMENT OF THE COLUMNS. Perhaps it may be thought that the INTENTION ought to ftand firft, and that the PROCESS and RESULT fhould be joined to each other. The contrary arrangement was adopted merely for the fake of *uniformity*; the Intention, or middle Column, being confidered as the *body of the Page*; the other two Columns as *explanatory margins*. And to the Author, to whom this arrangement is become familiar, any other difpofition would appear aukward. Befides, the PROCESS is frequently made, before any INTENTION with refpect to an *Experiment* is formed; and, indeed, every Agricultural Experimentalift will find this to be the *cheapeft* mode of Experimenting. If he make out a lift of *Experiments to be made*, and, *without waiting for favourable opportunities*, fet about making them, he will probably find the tafk in fome degree arduous: his men will, perhaps, be as aukward and difpleafed, as he will be embarraffed and out of humour. But if, when his Carter has committed a blunder, or has been guilty of a partial neglect, &c. &c. &c. he make a *Virtue of Neceffity*, and, inftead of reprobating the untowardnefs of the circumftance, *regifter it as a comparative Experiment*, Experimenting will become doubly ufeful.

In this incidental way, many of the following Experiments were made: As, No. 1. when *time* would not admit of the whole Field's being crofs-harrowed *before* fowing: No. 11. when the *feafon* would not permit (*a*) to be fown with beans; where *accident* formed a road; and a fhrubby bank *happened* to be grubbed: No. 21 where the mud happened not to cover the whole field: No. 67. where (*a*) was crofs-harrowed by a miftake of the Plowman, &c. &c. &c. Yet thefe in general are as decifive as thofe which coft the Writer many a ftep, and the Labourers who affifted in making them, many a half-fmothered imprecation.

Perhaps the moft expreffive way of *regiftering the identity of the place of Field-Experiments* is, in general, to draw a *diagram of the Field*, and identify the place of Experiment by the *number of Lands*, &c. examples of which mode of Regifter are given in Experiments No. 11 and 12.

A prin-

THE INTRODUCTION

A principal part of the following Experiments were regiftered in this diagrammatic manner; others were regiftered in *words* only, as No. 5, 9, 10, 15, 20, &c. which ftand as in the Autograph; others are defcribed by Gates, Trees, Hedges, &c. But as thefe details would be uninterefting to the Reader; and as a fimple letter is more perfpicuous, the diagrams and verbal defcriptions are fuppreffed in the publication; excepting thofe above-mentioned, which are publifhed as hints to thofe perfons who may wifh to acquire AGRICULTURAL KNOWLEDGE, by making COMPARATIVE EXPERIMENTS.

OBSERVING THE RESULTS OF EXPERIMENTS—is a very ferious Employment; yet if the Experiments are not too numerous, it is, to a philofophic Mind, a very pleafing one: it is receiving, in the Handwriting of Nature, an anfwer to a petition offered up at his throne.

The RESULT may fometimes be *decifive*, without being, to fuperficial infpection, *obvious*. As for inftance, in an Experiment on the *change of Seed-Wheat*—the quantity of *ftraw* and the length of the *ears* may be the fame; but on more minute infpection, the *grain* from one fpecies of feed may be large, bright and plump; while that from the other may be fmall, dull, and fhrivelly. Indeed, in an Experiment of this kind, nothing but actually thrafhing out an equal portion from each fpecies of feed, and *weighing* their feparate products, can be *nicely* determinate. This is an operation, however, which fteals away more attendance and attention both of mafter and fervant, than can, in common prudence, be fpared from the ordinary bufinefs of Harveft. Even meafuring out the ground, attending the cutting, and weighing the produce, *in the Field*, is more inimical to the neceffary Buftle of the Day, than any one who has not experienced it can be aware of. Befides, a Refult which does not, *to clofe and repeated infpections*, offer an *obvious difparity*, or at leaft a *perceptible difference* to a *judge*, is not *fufficiently decifive*. Nor is *one Refult*, be the difparity ever fo obvious, *fufficiently decifive*: it is only an *evidence* which muft be corroborated by a feries of fimilar *pofitive evidences*, before a *final judgement* can be paffed with fafety.

GENERALLY, There is nothing more dangerous to EXPERIMENTING than making *too many* Experiments; efpecially by a man who has the
immediate

immediate fuperintendance of a *large Farm*; as both his Farm and his Experiments are thereby in danger of being neglected. A few Subjects well chofen, carefully made, fully regiftered, and duly attended to, are far more valuable than a larger number in any degree neglected. In this light, a *fmall Farm* is the moft eligible; but a fmall Farm may want that *variety* which a *large Farm* may afford; and a *variegated Farm*, whether large or fmall, is evidently the moft proper for Experimenting. There is, however, fcarcely *any Farm* which will not, in the courfe of a few years, afford a variety of valuable Experiments: valuable not only to the Poffeffor of the Farm on which they are made, but to every man whofe foil and fituation is fimilar.

WITH RESPECT TO THE FOLLOWING EXPERIMENTS,—they are pub-lifhed as AN OVERTURE TO EXPERIMENTING, rather than as a COURSE OF EXPERIMENTS in themfelves fufficiently important to become the objects of publication. As EXPERIMENTS OF AGRICULTURE, however, it is hoped they will not be thought wholly unimportant; and the Writer thinks it proper, indeed neceffary, to aver, that to the beft of his knowledge they are *ftrictly authentic*. Had they been conti-nued with the fame attention and affiduity with which they were be-gun, they would have been far more numerous than they now are; but the Autumn of 1777 was engroffed by the Publication of the MINUTES OF AGRICULTURE; and the Spring of 1778 perplexed by a lefs agreeable circumftance: and a man who attends to the Proceffes of Experimenting, fhould have his head at leifure, and his heart at eafe.

Had the collection been larger, many of thofe now publifhed might with propriety have been fuppreffed; but the Author judged, that although an Experiment may be *undecifive* of the intention for which it was immediately made, it may neverthelefs convey fome ufeful information; and, in the prefent inftance, will at leaft convey a HINT to the NOVITIAL EXPERIMENTALIST.

Before the EXPERIMENTS themfelves be enumerated, it will be ne-ceffary to give a minute *Defcription of the Farm* on which they were made; as alfo an *explanation of the technical terms* made ufe of in the enumeration.

DESCRIP-

THE INTRODUCTION

DESCRIPTION of the FARM.

CLIMATURE. *Latitude*—51°. 24′. *Elevation*,—with refpect to the *fea*, confiderable; but with refpect to the *neighbouring country* the fituation is low; being over-looked by the hills of Norwood from the North, and by the mountainous, though cultivated, hills of Surry and Kent from the South and South-Eaft: with refpect to *itfelf*, the elevation is various, having a confiderable fwell towards each extremity. *Afpect*—various: the central parts nearly level: the extremities, fronting each other. The opening to the Weft, wide and expanfive; the view being terminated by the diftant hills of Hampftead and Harrow. The opening, or rather avenue (being contracted by Shooter's-Hill and the other inferior hills of Kent) to the North-Eaft is ftill more extenfive, being clofed by the very diftant high-lands of Effex. *Temperature of the air*,—mild, and what is perhaps erroneoufly termed moift. The *Seafons*, with refpect to *Harveft*,—ten days or a fortnight forwarder than thofe of the neighbouring hills of Surry and Kent, though fituated but three or four miles diftant from them.

LOCALITY. From ten to twelve miles South from the *Metropolis*; and from one to two miles Eaft from Croydon in Surry, a *Market-Town*. Adjoins to two *Commons*. Lies, with refpect to itfelf, *ftraggling* and difunited; and, in part, intermixed with the diminutive Patches of four or five diftinct and fcattered *Common-Fields*.

SOIL. Various.

SUBSOIL. Alfo various; as appears by the following *Defcription of the Fields*.

WATER. The central part very well watered; the more extreme parts, in general, ill watered.

ROAD. To fome parts of the Farm, very good; to other parts, bad. *No water-carriage*.

DESCRIP-

TO THE EXPERIMENTS.

DESCRIPTION OF THE FIELDS.

Names	Size	Soil	Subfoil	Afpect, &c.
A. 1.	5 Acres *	Clayey Loam	Retentive ‡	Very gently Eaftern
2.	3 ——	Tenacious Clayey Loam †	— —	Gently Southern
3.	$3\frac{1}{2}$ ——	— — — —	— —	Bold to the South
4.	4 ——	— — — —	— —	— —
5.	$1\frac{1}{2}$ ——	— — — —	— —	Eaftern ; but nearly level
B. 1.	$3\frac{1}{4}$ ——	— — — —	— —	— —
2.	$2\frac{3}{4}$ ——	— — — —	— —	Gently Northern
3.	$4\frac{1}{2}$ ——	— — — —	— —	— —
4.	$4\frac{1}{4}$ ——	— — — —	— —	— —
C. 1.	$2\frac{1}{4}$ ——	— — — —	— —	— —
2.	3 ——	Clayey Loam	— —	Northern ; but nearly level
D. 1.	$7\frac{3}{4}$ ——	— — — —	— —	Various ; principally Weftern
2.	2 ——	— — — —	— —	Gently Northern
3.	$1\frac{7}{8}$ ——	— — — —	— —	Level
E. 1.	$2\frac{7}{8}$ ——	Adhefive Clayey Loam	— —	Very gently Weftern
2.	$2\frac{5}{8}$ ——	— — — —	— —	— —
3.	$4\frac{3}{4}$ ——	Clayey Loam	— —	Gently Weftern
F. 1.	$4\frac{1}{4}$ ——	Black Moory Sandy Loam	Quick-fand	Eaftern ; but nearly level
2.	$5\frac{1}{2}$ ——	— — — —	— —	— —
G. 1.	$4\frac{1}{2}$ ——	Strong Sandy Loam	— —	— —
2.	$5\frac{3}{4}$ ——	— — — —	— —	— —
H. 1.	$3\frac{1}{2}$ ——	Spongey Sandy Loam	— —	— —
2.	$6\frac{1}{2}$ ——	Sandy Loam	— —	— —
I. 1.	$5\frac{1}{2}$ ——	Strong Sandy Loam	— —	— —
2.	$6\frac{1}{2}$ ——	Loam	Retentive	Gently Eaftern
K. 1.	$1\frac{7}{8}$ ——	Clayey Loam	— —	Level
2.	$5\frac{1}{4}$ ——	Mellow Clayey Loam	— —	Weftern ; but nearly level
3.	$11\frac{1}{2}$ ——	Tenacious Clayey Loam	Retentive	Nearly level ; inclined to nor.
4.	$6\frac{3}{8}$ ——	— — — —	— —	Very gently Weftern
L. 1.	$20\frac{1}{4}$ ——	Clayey Loam, Loam & Sandy Loam	— —	Very gently Northern
2.	$4\frac{3}{4}$ ——	Tenacious Clayey Loam	— —	Very gently Weftern
M. 1.	$5\frac{1}{8}$ ——	Loam and gravelly Loam	Various	Gently Northern
2.	$3\frac{1}{8}$ ——	Loam	Retentive	Very gently Weftern
3.	$4\frac{1}{8}$ ——	{ Rich gravelly Loam with boggy Patches	Various	Very gently Northern
4.	$3\frac{1}{2}$ ——		— —	Gently Northern
5.	$1\frac{1}{2}$ ——	Gravelly Loam	Abforbent	— —
6.	$4\frac{1}{2}$ ——	Gravelly and boggy	Various	— —
N. 1.	5 ——	Gravelly Loam, &c.	Abforbent	— —
2.	3 ——	Sandy grav Loam	— —	— —
3.	3 ——	Strong gravelly Loam	Various	Various
4.	3 ——	— — — —	— —	Various ; chiefly Southern
5.	$4\frac{3}{4}$ ——	Gravelly Loam	Springy	Various ; chiefly Northern
6.	4 ——	Gravel and gravelly Loam	Various	Gently Northern
7.	$1\frac{1}{8}$ ——	Strong gravelly Loam	Retentive	Bold to the South
	$196\frac{5}{8}$			

* The fractional Roods and Perches are omitted, as well for concifenefs as perfpicuity.

‡ Every Subfoil is confidered as *retentive*, which retains or checks the fuperfluous rain-water which drains thro· the Soil or Vegetative Stratum ; whether the Retention proceeds from the tenacity of the Subfoil itfelf, or from Retentivenefs of fome inferior Stratum, or from a partial Landfpring, &c.

† This is a peculiar fpecies of Soil : it is uncommonly adhefive and difficult to plow. It is particularly mentione a MINUTE of the 6th of February, 1776.

Names	Size	Soil	Subfoil	Afpect, &c.
	196 5/8 Acres			
O. 1.	13 1/2 ——	Gravelly Loam with boggy Patches	Various	Various; chiefly Northern
2.	2 ——	Sandy gravelly Loam	Abforbent	Gently Weftern
3.	2 3/4 ——	Part gravelly; part ftiff	Various	Part Northern; part Southern
P. 1.	4 ——	Gravelly Loam and boggy	— —	Gently Northern
2.	12 ——	Varied gravelly Loam	Abforbent	In general, full to the North
3.	4 ——	Gravelly ftrong Loam	Retentive	Gently Northern
R. 1.	2 3/8 ——	Gravelly fandy Loam	Abforbent	Very gently Northern
2.	1 ——	— — — —	— —	Gently Northern
3.	2 ——	— — — —	— —	— —
4.	0 7/8 ——	— — — —	— —	— —
5.	2 ——	— — — —	— —	— —
6.	1 3/4 ——	Gravelly ftrong Loam	Retentive	— —
S. 1.	1 1/2 ——	Spongey fandy Loam	Springy	Level
2.	1 1/2 ——	— — — —	— —	—
3.	1 1/2 ——	Spongey fandy Loam	Springy	Level
4.	1 1/2 ——	— — — —	— —	—
5.	1 1/2 ——	— — — —	— —	—
T. 1.	0 1/4 ——	Sandy Loam	— —	—
2.	1 1/4 ——	— — — —	— —	—
3.	{ 2 1/8 ——	— — — —	— —	Gently Eaftern
	{ 4 ——	Clayey Loam	Retentive	— —
4.	5 1/8 ——	— — — —	Various	Level
U.	4 ——	Sandy Loam	Abforbent	— —

269 1/8
21 7/8 Hedges, Roads, and Wafte.
291 including Fences, &c.

EXPLANATION

EXPLANATION

OF

TECHNICAL TERMS.

A

TO Acclivate. To make *acclivous:* to raife the Soil from *level* to *floping* ; fo as to fhed off fuperfluous Rain-water.

To Adjust. To put the Soil, after it has been fown and harrowed, into fuch a ftate as that the fuperfluous moifture may not only be fhed off with greater facility, but that it may be readily conveyed out of the Field ; and at the fame time to leave the Soil in a hufband-like manner.

B.

Burning. Liable to be fcorched by hot, droughty Weather.

C.

Cocklits. Small Cocks.

Cómpost. A compound of different Manures.

To Compóst. To mix or affimilate Manures.

Crop. This term is (agreeable to common acceptation) ufed ambiguoufly : it fignifies not only the *Quantity of Produce,* but is alfo a general name for any *Species of agricultural Vegetables.*

Cross-furrow. The *Gripe,* or *Water-furrow,* which receives the fuperfluous Rain-water from the Interfurrows, and conveys it into the Ditch or outlet.

D.

To Desposite. To *lay-up, lodge,* or *place* the Soil ; whether it be in *round Ridges, flat Beds,* or *entirely level.*

Dog-

EXPLANATION OF TERMS.

Dog-Days Fallow. A Fallow broke up and ftirred during the heat of Dog-Days; in oppofition to a *Winter*, a *Spring*, or a *Summer-Fallow*. It cannot with propriety be termed an *Autumnal-Fallow*; becaufe it is, or ought to be, broke-up in the height of *Summer*.

Dressing. A *Manuring*. This is a *Farrier-like* phrafe I have endeavoured to avoid.

Dung. See Note, page 111.

F.

Flutes. Seed-Seams formed by an Implement called a *Flute*. See Note (*d*) page 54, of the Digest of the Minutes of Agriculture.

Furrow. This vague term is divided into *Plow-furrow*; *Inter-furrow*;---*Crofs-furrow*;--and *Plit*.

H.

Haw. The *Head* or *Panicle* of the *Oat*.

Heart. The ftate of the Soil with refpect to Manure; or, more generally, with refpect to Vegetable food.

I.

Jag. A fmall Load.

Inter-furrow. The Trench or Drain left between two Beds or Ridges.

L.

Ley. *Grafs*-land; whether *annual* or *perennial*; in oppofition to land which is occupied by *Corn*, or by *Fallow*.

M.

To Make. When applied to *Hay*, is to *dry*; or rather to diffipate its fuperfluous Sap.

Melioration. See Note, page 51.

Mixgrass.

EXPLANATION OF TERMS.

MIXGRASS. Any mixture of graffes cultivated with an intention to produce a perennial Ley, or Meadow.

P.

PEABEANS. A mixture of *Peafe* and *Beans*.

PLIT. The portion of Soil which is turned in the operation of Plowing. See MINUTE of 13th of May, 1775.

PLOW-FURROW. The trench left by the Plow.

Q.

QUONDAL. A term made ufe of to defcribe what a Field *has recently been*. It is the characteriftic of the Field, or Soil, after the *Crop* is off. See MINUTE of 5th November, 1775.

R.

RIDGLET. A *diminutive Ridge*; more efpecially thofe raifed by TRENCHING.

S.

SEED-SEAMS. The Interftices between the Plits, as left by the Plow ; or thofe afterwards made by Fluting, Drilling, &c.

SODBURYING. Cutting off the foddy edge of the Plit, and burying it in the preceding Plow-furrow.

SUBSOIL. The ftratum of earth which lies immediately under the cultivated Soil ; and is, as it were, the platform on which the vegetative ftratum undergoes the operations of Hufbandry.

To SUBPLOW. To run a *Share* through, or below, the Soil, without *turning* it.

SUCCESSION. See page 168.

T.

TARE-BARLEY. A mixture of *Tares* and *Barley*.

To TILLER. To fpread, branch-out, or ramify.

TILTH.

EXPLANATION OF TERMS.

TILTH. The state of the Soil with respect to Tillage.

TOP-DRESSING. Manure laid on the *Crop*.

TRENCHING. Cutting the Soil into *Trenches* with the Plow, leaving narrow Ridglits between them; in order to expose the Soil to the action of the Atmosphere

V.

VEGETIZING-PROCESS. See page 183.

VERDAGE. *Green* Herbage; in opposition to *Hay*, *Straw*, or other *dry* Herbage : more especially the *Green Meat* given to Horses or Cattle in the stable—barbarously called *Soiling*.

W.

WAD. A small bundle of Hay or Corn. See Note, p. 92.

WHIP-REIN-PLOW. A small Plow drawn by two Oxen or Horses, which are guided and driven by hempen *Reins*, made in such a manner as to answer the purpose of a *Whip*.

WINTER-PROUD. See Note to *Experiment* 14.

Y.

YIELD. The *actual Produce*; particularly *the Quantity of Grain* after it has been separated from the *Straw*.

SYSTEMATICAL INDEX

TO THE

OBSERVATIONS.

AGRICULTURE.

Elements

  Earth *
- Farms - - - - 109
- Soils - - - - 110
- Manures - - - 111

  Vegetables
- Wheat - - - 45 & 95
- Barley - - - - - 26—105
- Oats - - - - 65—101
- Tare-Barley - - - 30
- Peafe - - - 89
- Tares - - - - 63
- Peabeans - - - 34
- Clover - - - - 16— 75
- Meadow - - - 22— 85
- Mixgrafs - - - - 39— 80

  Animals ‡
- Horfes - - - ⎫
- Cattle - - - - ⎪
- Sheep - - - ⎬ 187
- Hogs - - - ⎪
- Poultry - - ⎪
- Bees - - - ⎭

Agents

  Nature
- Animal Economy - - ⎫
- Vegetation - - ⎬ 114
- Weather - - - ⎭

  Art
- Servants - - - 165
- Beafts of Labour - - ibid.
- Implements - - - 166

Proceffes

  Vegetable Management
- Divifion of Farms - - 167
- Succeffion - - - 168
- Soil-Procefs - - - 170
- Manure-Procefs - - 111
- Seed-Procefs - - 171
- Vegetizing-Procefs - - 183
- Vegetable-Procefs - - 114
- Farm-Yard Management - ⎫
- Markets - - - ⎬ 185
- Accounts - - - ⎭

  Animal Management
- Breeding - - ⎫
- Rearing - - - ⎪
- Dairying - - ⎬ 187
- Fatting - - - ⎪
- Markets - - - ⎪
- Accounts - - - ⎭

* See Note, page 114.   ‡ See page 114.

ERRORS of the PRESS.

P. 14. 1. 9. from Bottom, for 21 read 21$\frac{7}{8}$.

— 22. 1. 6. ——————, for *are* read *were*.

— 24. 1. 12. —— Top, for *unmowed* read *unmoved*.

— 93. Note, 1. 2. from Bottom, for *reaps* read *reap*.

EXPERIMENTS

CONCERNING

AGRICULTURE.

The Process.	The Intention.	The Result,
———April 1776.———	———No. I.———	——— July 1777. ———
Sowed the further fide *before,* the hither fide *after* crofs-harrowing, The whole was afterwards rolled.	(In N. 5.) SOWING CLOVER. Should Clover Seed be buried deep, or fhould it be merely covered? *Either:* by this Experiment.	The whole field a very *even,* good crop.
——— April '76. ———	———No. II.———	——— July '78. ———
a, light-rolled. *b,* heavy-rolled, immediately after the feed was covered. *c,* not rolled.	(In N. 6.) ROLLING GRAVEL. Should a fharp, gravelly loam be condenfed? or fhould it be left porous? *Undecifive.*	Neither the *Barley* nor the *Clover* received any perceptible advantage or detriment by the rolling.

A

THE

EXPERIMENTS OF AGRICULTURE.

THE PROCESS.	THE INTENTION.	THE RESULT.
——— April '76. ———	——— No. III. ———	——— Aug, '77. ———
(*a*) fown before fpikey-rolling, (*b*) fown after it. The whole field was afterwards fwept with the bufh-harrow.	(In N. 4.) SOWING MIX-GRASS. Should the feed be deeply or flightly covered? *Either:* by this Experiment.	This field was paftured. Not the fmalleft difference to be obferved; excepting that where the tilth was fine, the plants are thick; where cloddy, thin.
——— April '76. ———	——— No, IV. ———	——— July '77. ———
Harrowed all the Wheat of I. 2, with a pair of very light harrows, in order to raife frefh mould for the feed to drop upon; except a belt acrofs the middle.	(In I. 2.) SOWING CLOVER. Is harrowing the foil before fowing Clover over Wheat beneficial to the crop? *No:* not by this Experiment.	The whole field was *equally* good. *N. B.* This belt was neither harrowed before, nor rolled after fowing: H. 1. the fame, and very good.
——— Sept, '76. ———	——— No. V. ———	——— Aug. '77. ———
Two lands againft the road, from the winding part of the road upwards, *dry*; the reft *pickled*, in lime-water brine.	(In P. 1.) SOWING WHEAT. Is brineing the feed advantageous to the crop? *Not by this Experiment.*	It is remarkable that thefe two lands are *forwarder* and a *better* crop than the reft of the field; and as totally free from *fmut.*
——— Oct. '76. ———	——— No. VI. ———	——— Aug. '77. ———
a, a, fown with feed raifed on a fharp gravel (an oppofite foil). *b,* with the very fame fpecies, raifed on a clayey loam (a fimilar foil).	(In L. 1.) SEED WHEAT. Is changing the feed from foil to foil of different fpecies, beneficial to the crop? *No.*	Remarkable! *b,* was always the *rankeft* crop; and is now *more lodged* than *a,* and feveral days *forwarder.*

EXPERIMENTS OF AGRICULTURE.

THE PROCESS.	THE INTENTION.	THE RESULT.
—— Oct. '76. ——	—— No. VII. ——	—— Aug. '77. ——
a, a, pickled in strong lime-water, salted until it would bear an egg. *b,* sown dry.	(In L. 1.) SOWING WHEAT. Is brincing the seed beneficial? *No.*	The crops equal; and not a smutty ear in the whole field: at least, not in the part sown dry.
—— 1776. ——	—— No. VIII. ——	—— Aug '77. ——
a, a Summer-Fallow of six plowings (including the breaking-up). *b,* Mazagan Beans in drills — horse-hoed, hand-hoed, hand-weeded, and, (with the spring-plowing) five times plowed.	(In L. 2.) FALLOWING. Is a Summer-Fallow or a Fallow-Crop more advantageous to the succeeding crops? *A Summer-Fallow *.*	(*a*) much the rankest crop of *Wheat,* with a gardenly quondal. **Aug '78.** *a,* obviously the best crop of *Oats,* and by much the cleanest quondal.
—— Oct. '76. ——	—— No. IX. ——	—— Aug. '77. ——
17 lands over. 2 —— under. 4 —— over. 2 —— under. 66 —— over. The quantities of feed equal: about 2½ bushels an acre.	(In L. 1.) SOWING WHEAT. Is under-plit or over-plit preferable? *Over-plit;* with the same quantity of feed.	The under-plit the stronger straw and the larger ears; but much the thinnest, much the foulest, and much the shabbiest crop.
—— Oct. '76. ——	—— No. X. ——	—— Aug. '77. ——
The outsides of the two *double* lands abovementioned were harrowed thrice in a place: the insides left rough..	(In L. 1.) SOWING WHEAT. Is it better to harrow after sowing *under,* or to leave the soil in rough plit? *Undecisive.*	Equally thin, and equally foul. But perhaps (indeed most like) the outsides were not sown so thick as the insides.

* Whenever a positive answer is given, " by this Experiment" must always be *understood.* For it is not *one* Result; but a *series* of *similar* Results which amount to certainty. (See MINUTES of AGRICULTURE, 7. Nov. 1776)

EXPERIMENTS OF AGRICULTURE.

THE PROCESS.	THE INTENTION.	THE RESULT.

———— 1776. ———— | **———— No. XI. ————** | **———— Aug. '77. ————**

a, a,   Tare-barley.
b, b,   Summer-fallow.
c,   Beans.
d,   a beaten road *.
e,   a grubbed bank.

(In L. 1.)

SUCCESSION.

What does Wheat like to
fucceed?
A Summer-Fallow ;
   or
A beaten road ;
   or
Virgin Earth ;
   or
Tare-Barley ;
   or
Beans.

b,   the rankeft crop,
d,   the ftrongeft.
e,   very good.
a & c, equally good ; and
equally foul, compared with
the Summer-Fallow.

N. B.   The whole well
dunged ; except (e).

———— Oct. '76. ———— | **———— No. XII. ————** | **———— Aug. '77. ————**

a, 1. (Ta. Bar.) had 7
a, 2. (Ta. Bar.) — 6
b,   (Ta. Bar.) — 5
c,   (Road)   — 5
d, (Sum. Fal.) — 6 } plowings.
e, (Grub. B.) — 7
f, (Sum. Fal.) — 7
g,   (Beans) — 4
h,   (Beans) — 5
(including the breaking-up).

(In L. 1.)

FALLOWING.

How does the quantity of
Tillage affect the fucceeding
crops and quondals?
*The crops good, and the
quondals clean in proportion
to the number of plowings ;
except that a Summer-Fallow
of fix plowings is preferable
to a Fallow-Crop with the
fame quantity of tillage.*

(Wheat.)

d & f, the rankeft crop,
and cleaneft quondal.
c, the ftrongeft, but foul.
a 1 & b (croffed) much
better than a 2 & g, (not
croffed).
g, weedier than h.
h, weedy.
e, a good crop, tho' un-
dunged.

Aug. '78.

(Oats.)

c, d, e, and f, much the
beft and cleaneft crop; and
the quondal beyond compa-
rifon better than the parts
fallow-cropped.

---

* Made by the plow-teams and dung-carts.

EXPERIMENTS OF AGRICULTURE.

THE PROCESS.	THE INTENTION.	THE RESULT.
——— Nov. '76. ———	——— No. XIII. ———	——— Aug. '77. ———
a, was plowed the 18 *Sep.* *b*, the 23 *Oct.* The whole fluted and sown promiscuously the 2 *Nov.*	(In P. 3.) FLUTING. Is it better to flute the fresh or the stale plit of a Clover-Ley for Wheat on a gravelly loam ? *The stale plit.*	*a*, is, very perceptibly, the strongest crops, and the cleanest quondal.
——— Nov. '76. ———	——— No. XIV. ———	——— Aug. '77. ———
The whole of the ley part meliorated with 15 jags of compost an acre; after plowing, but before fluting. The soil otherwise out of heart. No comparative Experiment made.	(In P. 3.) MANURING. What effect has top-dressing before fluting a strong loam for Wheat? *The effect not equal to expectation:* it may, perhaps, answer better on a lighter soil.	The crop about thirty shocks (of 10 sheaves) an acre: the straw short, and the ears small. During winter and spring, it promised a better crop.—But perhaps the manure lying near the surface, and the soil being too tenacious to let it down, the crop was rendered *winter-proud**.
——— Nov. '76. ———	——— No. XV. ———	——— Aug. '77. ———
There are several *obvious* Experiments in P. 2, both on the gravelly and on the springy parts.	(In P 2.) DEPOSITING GRAVEL. Should a sharp gravelly loam be laid up round, or should it be plowed flat ? *It should be landed up.*	It is remarkable that the round lands, every where, not only looked best during winter; but are now obviously the best crop

* This is a Provincial Phrase very expressive of the idea it represents: it is applied to Wheat, which in Winter puts on a more splendid appearance than that which it can support thro' the ensuing Summer. The Soil having exerted itself too freely on the infant plants, the ramifications are become too numerous to be reared to maturity: The crop, in the Spring, consequently declines, and at Harvest falls short of its neighbour, which in Winter put on a more *humble* appearance.

4

THE

EXPERIMENTS OF AGRICULTURE.

THE PROCESS.	THE INTENTION.	THE RESULT.
——— Nov. '76. ———	——— No. XVI. ———	——— Aug. '77. ———
a, A bushel and a half. *b*, Four bushels. *c*, Two and a half bushels an acre.	(In P. 2.) SOWING WHEAT. Is thick or thin sowing preferable on a gravel ? *Thin,* if the summer prove *wet.*	*a*, Much the best crop. *c*, About 50 shocks an acre *b*, Very bad : scarcely worth reaping ! The straw and the ears equally short and puny ; *this very wet year.*
——— Nov. '76. ———	———No. XVII.———	——— Aug. '77. ———
a, a, a, Compost of Dung and Mould. *b, b,* Good ripe Dung. The quantities equal : about 15 jags an acre.	(In P. 2.) MANURE. Is dung or compost preferable as a top-dressing, before fluting a retentive loam for Wheat ? *They are equally useless.*	No perceptible difference. The whole a thin, straggling crop ; although in winter it looked very promising.———See Experiment, No. XIV
——— Nov. '76. ———	———No. XVIII. ———	——— Aug. '77. ———
a, a, a, raised on a gravelly loam (a similar soil). *b, b,* on a clayey loam (an opposite soil), the species exactly the same ; being two years ago produced from one stock.	(In P. 2.) SEED WHEAT. Is it advantageous to change it from soil to soil of different species ? *No :* at least, not by this Experiment.	*a, a, a,* the forwardest, and the most lodged.———The straw punier and more blighted than *b, b.* The crops, on the whole, equal ; or as nearly so as can be judged this awkward year for experimenting.
——— Nov. '76. ———	———No. XIX. ———	——— Aug. '77. ———
a, a, sown dry The rest of the division pickled in Lime - Water - Brine.	(In P. 2.) SOWING WHEAT. Does any benefit arise from brineing and limeing the seed ? *No :* none at all.	The crops equal, and equally free from smut.

EXPERIMENTS OF AGRICULTURE.

THE PROCESS.	THE INTENTION.	THE RESULT.
——— Nov. '76 ———	—— No. XX. ——	——— Jan. '77. ——
Sowed about half-an-acre in Garden-field, the 7 *Nov.* Fluted one fide of each ridge, and left the other fide in whole-plit. Two lands next to the wheat fown dry.—The four middle ones pickled in lime-water brine.—The two hither ones, and the fruftum, dry.	(In S. 2.) SOWING BARLEY. Is fowing it in November eligible? *Yes.* Should it be fown, in Autumn, over Flutes, or over the fresh Plit? *Over Flutes.* Should the feed be pickled, or fhould it be fown dry? *Sown dry.*	The late fevere weather has made the whole hang its head, and look very wan: fome of the tips quite nipped with the froft. Aug. '77. The whole a good crop: two jags an acre; and the ftraw comes very timely for thatch. The *Flutes*, the *eveneft* crop The *dry*, equally as good as the *prepared*.
——— Nov. '76. ———	—— No. XXI. ——	——— July '77. ———
a, Sixteen jags an acre of pond-mud, caft out about 15 months ago; no mixture whatever. *b*, Nothing.	(In D. 2.) MANURE. Are Pond-Dregs, laid on *in November*, ferviceable to a clayey meadow? *Not obviouſly.*	The whole a good crop: no obvious difference this year. July '78. Nor this: the whole a good crop, and remarkably full of white Clover, which I attributed to the mud, until I perceived the part not mudded is the fame.
——— Dec. '76. ———	—— No. XXII. ——	——— Aug '77. ——
a, Ten loads of yard-dung, (made from ftraw and fern, with fome fhovellings of dirt) an acre. *b*, Not dunged.	(In A. 3.) MANURING. Is it eligible to lay Dung on young graffes on a clayey loam, *in December?* *No.*	The whole a very bad crop: not worth mowing, (See Exp. No. LVI.) tho' a Hay-Year. The part dunged is difcernible,—and that is all! Aug. '78 The fame may be faid this year.

THE

EXPERIMENTS OF AGRICULTURE.

THE PROCESS.	THE INTENTION.	THE RESULT.
——— Dec. '76. ———	——— No. XXIII. ———	——— Aug. '77. ———
a, a, a, About 24 jags of weak compoſt (about 2-3ds mould and 1-3d dung) an acre. *b, b, b,* Nothing.	(In C. 1.) **MANURING.** Of what uſe is Compoſt, laid on *in December,* to young graſſes on a clayey loam ? *Of none.*	Aſtoniſhing! Not the trace of a difference: the whole being bad ; though a Hay-Year. Aug. '78. Exactly the ſame ! The whole a *very* bad crop : The Compoſt *wholly* unſerviceable !
——— Dec. '76. ———	——— No. XXIV. ———	———June '77.———
a, Saw-duſt and dung, half-and-half, well digeſted and thoroughly compoſted. *b,* Clay, night-ſoil, loam, dung, &c. The quantities equal : more than 20 jags an acre *laid on* the young ſeeds.	(In I. 1.) **MANURE.** Has Saw-Duſt Compoſt or Loam-Compoſt the preference as a Dreſſing for Clover and Wheat ? *Undeciſive* *.	No difference, the firſt crop of clover. Aug. '77. Nor, the ſecond crop. Aug. '78. Nor, the ſucceeding crop of wheat.
——— Jan. '77. ———	——— No. XXV. ———	——— July '77. ———
a, a, a, Sandy loam and dung : about 2-3ds dung : *b, b, b,* Pond-mud and dung : about 1-3d dung.	(In I. 2.) **MANURE.** Is Pond-Dreg Compoſt or Sandy Loam Compoſt preferable as a dreſſing for Clover and Wheat on a clayey loam ? *Dubious.*	Equally unſerviceable to the firſt crop of clover. Aug. '77. As likewiſe to the ſecond. Aug. '78. The wheat of *b,* if of either, is the larger crop ; but that of *a,* is forwarder.

* This Experiment does not, by any means, depreciate ſaw-duſt as a manure: perhaps if it were *plowed-in,* it would be equal to many other vegetable ſubſtances.

THE

EXPERIMENTS OF AGRICULTURE.

THE PROCESS.	THE INTENTION.	THE RESULT
—— Jan. '77. ——	—— No. XXVI. ——	—— July '77. ——
See laſt Experiment. The quantity, upwards of 20 jags an acre. *c*, Left wholly undreſſed.	(In I. 2.) **MANURING.** Is it proper to lay Compoſt on the ſurface of a clayey loam for Clover, *in January?* *No: exceedingly improper.*	The Compoſt of no uſe to the firſt crop. Aug. '77. Nor to the ſecond.
—— Feb. '77. ——	—— No. XXVII. ——	—— June '77. ——
a, a, Mud and 1-3d Dung. *b, b,* Mould and 2-3ds Dung. *c,* Undreſſed.	(In I. 2.) **MANURE.** Is Mud or Mould preferable to mix with Dung? *This Experiment is not deciſive.*	The crops of Clover from *a, b,* and *c* are *equally* good: neither of the Compoſts therefore acted; conſequently their comparative merits cannot be aſcertained by the *Clover.* Aug. '78. Nor by the *Wheat.*
—— Feb. '77. ——	—— No. XXVIII. ——	—— June '77. ——
See the laſt Experiment. The quantity upwards of 20 jags an acre.	(In I. 2.) **MANURING.** Is it, or is it not, good management to dreſs young Clover on a clayey loam, in *February?* *It is very bad management.*	*c* Full as good a crop as either *a,* or *b,* and obviouſly leſs graſſy. Aug. '77. The ſecond crop, the very ſame. Aug. '78. The difference between 20 jags of compoſt and nothing, is ſcarcely diſcernible!

B

THE

EXPERIMENTS OF AGRICULTURE.

The Process.	The Intention.	The Result.
——— Feb. '77.———	—— No. XXIX. ——	——— June '78. ———
The whole plaſhed rough, and the vacancies filled-up with Spray Boughs; except one rod in length, which was trimmed before plaſhing, and the vacancies ſupplied by naked rods. The whole afterwards trimmed cloſe on both ſides.	(Between S. and H.) **PLASHING HEDGES.** Is it better to trim the Plaſhers and fill-in with Rods, or to leave the Spray on, and fill-in with Spray Buſhes? *Trim the Plaſhers, and fill-in with Rods.*	The part trimmed and filled in with rods is much the beſt fence. *A live roddle hedge* is the moſt *perfect* fence on a farm. To put dead ſprayey boughs into a plaſhed hedge is very bad management. They retain the wet; hinder the circulation of the air; and prevent the plaſhers from germinating.
——— Feb. '77. ———	——— No. XXX. ———	——— Feb. '76. ———
The whole field buſh-harrowed, and the *wheat-ſtubble* ox-raked into the inter-furrows, and carted-off for yard-litter. *a,* Was afterward heavy-harrowed, and the weeds, &c. which the harrows eradicated, were oxraked into the inter-furrows, and there buried *. *b,* Was harrowed, but not afterwards raked. *c,* Only buſh-harrowed, and the ſtubble raked-off.	(In M. 6.) **DISCUMBERING.** What is the advantage of clearing the ſurface of light-land Quondals before plowing? *It is of no advantage to the immediate crop; but cleanſes the ſoil for future crops.* *It renders a gravelly loam too fallowy for peabeans; and perhaps generally it is not worth the labour attending it.*	*a,* Fluted the beſt. Feb. '77. The whole broke up ſo exceedingly fallowy, that one-third of the *beans* were left uncovered for want of ſeed-ſeams to receive them. April '77. An adjoining field totally unprepared, and which broke up in whole plit, has many more plants from the ſame quantity of ſeed. Aug. '77. The crops equal. *a,* Obviouſly the cleaneſt quondal. *b* and *c* (equally) graſſy when compared with *a.*

* The Quondal being tolerably clean, this left the ſurface almoſt as free as a fallow.

THE

THE PROCESS.	THE INTENTION.	THE RESULT.
——— Feb. '77. ———	———No. XXXI.———	——— Aug. '77. ———
a, Left rough. *b*, Bufh-harrowed. *c*, Bufh-harrowed, and oxraked; a row of ftubble being left, and afterwards buried by the plow, in every fecond inter-furrow. The lands are now reverfed, and confequently every alternate land has now a row of ftubble buried under its ridge; but the intervening lands are wholly deftubbled.	(In M. 4.) UNSTUBBLING. Is it better to take Wheat-ftubble off, or to leave it on gravelly loam? *Undecifive.* And is it better to carry it away when raked into rows, or to bury it in the inter-furrows? *It is very immaterial.*	(Peabeans.) The wet of this fummer has forced every weed, and has beaten the crop flat to the ground, fo that it is impoffible to fay pofitively which part is the cleaneft, or which is the beft crop. The latter part of the experiment, however, is fufficiently decifive; there being no perceptible difference between the lands in which ftubble is buried, and thofe in which there is none.
——— Feb. '77. ———	——— No. XXXII. ———	——— Aug. '77. ———
a, a, Fluted. *b,* Sown ower the frefh plits; which in general were plowed too narrow.	(In M. 5. and 6.) SOWING PEABEANS. Should Peabeans be fown over flutes, or over the frefh plit? *Over flutes gives the eveneft crop.* *Over the plit is more expeditious, and lefs expenfive.*	*a, a,* Perceptibly the eveneft, largeft and cleaneft crop. But whether the advantage be fuperior to the extraordinary labour, attendance and attention, needs further experiments.
——— Feb. '77. ———	———No. XXXIII.———	——— Aug. '77. ———
a, Marlbro' grey peafe, alone. *b,* Mazagan beans, alone *c,* Marlbro' greys and mazagans mixt, half-and-half. The quantities four bufhels an acre; except of the peafe, which was fomething lefs.	(In M. 4.) PEASE and BEANS. Are *Peafe, Beans,* or *Peabeans* moft eligible for a gravelly loam? *Peafe.*	*a,* (if either), the moft beaten down, and the cleaneft. *b,* Much the fouleft; and, where the gravel is fharp, very thin and fhort. *c,* The Peafe good; the Beans very ftraggling and puny: the Seed-beans were in a great meafure thrown away *.

* And by Obfervations laft year, Clean Peafe yielded full as well as Peabeans.

 THE

THE PROCESS.	THE INTENTION.	THE RESULT,
—— Feb. '77 ——	—— No. XXXIV. ——	—— Aug. '77. ——
a, a, Rolled very hard with a heavy roller, both before and after the laſt harrowing. *b,* Entirely unrolled.	(In M. 6.) ROLLING GRAVEL. Should a gravelly loam be rolled immediately after ſowing Peabeans ? *Dubious.*	*a, a,* Perceptibly the beſt. This reſult, however, may not always follow : the wet of this ſummer prevented the Gravel from *binding* *.
—— Mar: '77. ——	—— No. XXXV. ——	—— July. '77. ——
a, Whole. *b,* Broken by one tine of the ſmall round harrows, to prevent the ſeed from being buried. *c,* Fluted.	(In M. 1.) S O W I N G T A R E - B A R L E Y. Is it better to ſow Tarebarley on *gravelly loam,* over the freſh plit, *whole* ; or over the freſh plit, *broken* ; or over the freſh plit, *fluted ?* *Fluted, if either.*	The fluted the eveneſt crop. The harrowing before ſowing had no apparent effect.
—— Mar. '77. ——	—— No. XXXVI. ——	—— July '77. ——
a, Broken, by once in a place with the middle round harrows. The reſt, *whole.*	(In M. 1.) S O W I N G T A R E - B A R L E Y. Is it better to ſow it on a *ſtrong loam* over the *whole,* or over the *broken* plit ? *The whole plit.*	No poſitive difference ; if either, the *whole* is the beſt crop.

* *Gravel* is generally conſidered as a *light* ſoil, and treated accordingly : While, perhaps, a *gravel-walk,* inſtead of being fertiliſed, is rendered ſterile by *rolling.*

EXPERIMENTS OF AGRICULTURE.

THE PROCESS.	THE INTENTION.	THE RESULT.
—— Mar. '77. ——	—— No. XXXVII. ——	—— June '77. ——
The Tares of S. 5. *a*, Liquored—*b*, not. The Clover of H. 1. *a*, Liquored—*b*, not. The Clover of H. 2. *a*, Liquored—*b*, not. The liquor was of middling ftrength : it was very high-coloured and *foul*, but not *puddly*; and was carried on in wet weather.	(In S. 5.) (And H. 1.) (And H. 2.) YARD LIQUOR. Are 20 pipes an acre obvioufly ferviceable to Tares or Clover on fandy loam ? *No.*	The Tares of S. 5.—Of no perceptible advantage: The Clover of H. 1.—Of no advantage to the 1ft crop. The Clover of H. 2.— The fervice perceptible, but not obvious. Aug. '77. Of no fervice to the 2d crop of Clover. Aug. '78. Nor to the fucceeding crop of *Peafe*.
—— Mar. '77. ——	—— No. XXXVIII. ——	—— Aug. '77. ——
a, Scotch Oats, immediately from Alemouth, (on the borders of Scotland). *b*, Scotch Oats, raifed on the chalky hills of Surry. *c*, Poland Oats, raifed in E, a clayey loam.	(In B. 3. and 4.) SPECIES OF OATS. Which Species has the preference on a clayey loam ? *The Scotch, raifed on a chalky loam in England, are preferable to the fame fpecies, brought immediately from Scotland.*	*a*; Has given an immenfe crop of ftraw, but the grain is uncommonly thin. *c*, A fhabby-looking crop on the ground, but the grain is fine. *b*, A due proportion of ftraw and corn.
—— April '77. ——	—— No. XXXIX. ——	—— Aug. '77. ——
See Experiment, No. 34. *b*; Still left unrolled. Two belts of *a*, *a*, again heavy-rolled.	(In M. 6.) ROLLING GRAVEL. Is it, or is it not beneficial ? *Dubious.*	This additional rolling had no perceptible effect.——

EXPERIMENTS OF AGRICULTURE.

THE PROCESS.	THE INTENTION.	THE RESULT.
——— April '77. ———	——— No. XL. ———	——— Aug. '77. ———
a, a, Rolled, prefently after the crop was up. *b,* Not rolled. The Beans were in broad leaf; but had not run much to ftalk.	(In M. 6.) ROLLING PEABEANS. May they be rolled with a heavy roller, after they are up, with fafety? *Yes.*	*a,* If either, the beft crop. There is no *danger,* there-fore, in rolling Peafe nor Beans after they are up: and (*a*) was much better to mow.
——— April '77. ———	——— No. XLI. ———	——— July '77. ———
See Experiment, No. 2 A belt of (*a*), a belt of (*b*), and a belt of (*c*), of that Experiment, are now rolled with the heavy roller.	(In N. 6.) ROLLING CLOVER ON GRAVEL. Is rolling Clover on a fharp gravelly loam benefi-cial to the crop? *No:* it rather injures it.	(Firft Crop.) The belts, which were rolled in April, are percep-tibly, tho' not obvioufly, the worft Clover in the field. Aug. '77. (Second Crop.) No difference.
——— April '77. ———	——— No. XLII. ———	——— Aug. '77. ———
Rolled a diagonal belt from *a* to *b* with the heavy roller. The reft of the field not rolled.	(In L. 1.) ROLLING WHEAT ON CLAY. Should Wheat on a clayey loam be rolled in April? *No.*	Not the fmalleft difference to be traced; and the roll-ing, of courfe, loft labour.
——— 29. April '77. ———	——— No. XLIII. ———	——— Aug. '77. ———
Mowed part of a ridge in the rankeft part of the field, and gave the herbage to the oxen.—The plants a foot high.—Worth (at this time) three guineas an acre.	(In L. 1.) VEGETIZING WHEAT. If Wheat be very rank in the fpring, is it eligible to check it by cutting it for foiling? *Yes,* perhaps, if the win-ter has been dry *.	The part mowed is lefs lodged; a few days later; much fouler; but on the whole, a better crop, *this wet year,* than the parts not checked; *befides the* SPRING FEED.

* This, however muft not be taken as an infallible guide; for although the winter of 76-77 was dry, and the enfuing fummer proved wet; yet the winter of 77-78 was equally dry, and the drought continued through the fummer.

EXPERIMENTS OF AGRICULTURE.

THE PROCESS.	THE INTENTION.	THE RESULT.
——— April '77. ———	——— No. XLIV.———	——— Aug. '77. ———
a, a, Common Red Clover, raifed on a chalky loam. *b, b,* Cow-grafs, bought of Gordon. Both fown the fame day, by the fame men; the quantities equal; the clover feed the beft fample.	(In B. 3 and 4.) SPECIES OF CLOVER. Is there any difference between the common *Red Clover* and *Cowgrafs. Yes.* Which of them is preferable? *Cow-grafs,* for one crop, and after-pafture. *Clover,* if two crops of hay are wanted.	Both miffed in B. 3. June '78. In B 4.—The fpecies are obvioufly diftinct: the Clover blows three weeks before the Cowgrafs; has more *pith* and lefs *wood*; is not fo fweet to the tafte, nor fo large a crop as the Cowgrafs. Sept. '78. The fecond crop of Clover ftood for feed, the Cowgrafs was cut for *Hay in the middle of September.*
——— April, '77. ———	——— No. XLV. ———	——— June '78. ———
(In A. 4.) *a,* Mixgrafs and Cowgrafs. *b,* Mixgrafs, Cowgrafs and Ryegrafs. (In A. 5.) *a,* Ryegrafs and Cowgrafs. *b,* Ryegrafs and Mixgrafs. *c,* Mixgrafs and Cowgrafs. *d,* Mixgrafs (a mixture of Ribgrafs, Trefoil and white Clover) alone. The quantities equal.— About 14lb. an acre, all hand-raked in.	(In A 4 and 5.) LEY-GRASSES. Are Ryegrafs and Cowgrafs, Or, Ryegrafs and Mixgrafs, Or, Mixgrafs and Cowgrafs, Or, Mixgrafs alone, the moft eligible Graffes for leying a clayey loam with? *Mixgrafs and Cowgrafs, the firft year.*	(In A. 4.) *a,* The beft crop. (In A. 5.) *c,* The beft. *a,* The next. *d,* The next. *b,* Very bad—*The firft year.* It will be *five or fix years* before their comparative merits, as perennial LEY-GRASSES, can be afcertained.
——— April '77. ———	——— No. XLVI. ———	——— July '78. ———
Two lands in the middle of the field, fown with a mixture of White Clover, Ribgrafs, Trefoil and Cowgrafs.	(In A. 1.) RE-SOWING LEYS. If the feeds of ley-graffes have miffed the firft year, is it of any fervice to re-fow the fecond? *Not decifive.*	Not the leaft benefit received from the re-fowing. But the furface is foul; and this is not a decifive experiment.

4

THE

EXPERIMENTS OF AGRICULTURE.

THE PROCESS.	THE INTENTION.	THE RESULT.
——— April '77. ———	—— No. XLVII. ——	—— Oct. '77. ——
a, Red bunch hog Pota-toe. *b*, The common White Kidney. Part of the plants were cut from fmall, part from large Potatoes. Part were planted under, part upon the dung.	(In S. 3.) POTATOES. What comparifon do the *Red Bunch* and the *White Kidney* bear to each other? *The Red Bunch give a greater quantity. The White Kidney a finer quality.*	*a*, Much the largeft crop; but the Potatoes are of a coarfe, harfh quality. *b*, A much fmaller crop; but their quality is incomparably fuperior. Neither the *Plants*, nor the *Planting* were fufficiently attended to.
——— June '77. ———	—— No. XLVIII. ——	—— Aug. '78. ——
Plowed in with a buryfod plow, the Tare-barley and Ketlock of Clays (the Ketlock beginning to pod), except two lands of No. 16. (2½ lands from the Male-bank) which were mown for foiling, and plowed at the fame time with the other lands. The Herbage carried off worth about 40 s. an acre.	(In M. 1.) FOUL CROPS. Should Tare-barley, which is too full of ketlock to be fuffered to ftand for hay, be cut and carried off as green fodder, or fhould it be plowed in as manure? *Undecifive.*	(*Barley.*) It is very remarkable, that the 2¼ lands (next the Male-bank) *buried*, and the two lands *carried off*, are perfectly equal: but what is ftill more furprifing, the remainder of the fame piece, *buried*, is obvioufly worfe than the two lands (to which it joins) *carried off*. The parity of the one, and the difparity of the other, are evident to common obfervation.
——— July '77. ———	—— No. XLIX. ——	—— Aug. '78. ——
a, a, Cut for foiling. *b*, Plowed in with a buryfod plow. *a*, Worth about 50 s. an acre: *b*, about 40 s.	(In M. 2.) FOUL CROPS. Is it better management to *verdage* or *bury* a crop which is too foul to ftand? *Cut it for verdage.*	(*Barley*) Part of *a* is dubioufly worfe than *b*; but in other places, the part *verdaged*, and the part *buried*, are equal.

THE

EXPERIMENTS OF AGRICULTURE.

THE PROCESS.	THE INTENTION.	THE RESULT.
——— July '77. ———	——— No. L. ———	——— June '78. ———
The Clover of Barn-Field was out more than three weeks of rainy weather.—Made it into two ſtacks of equal ſize and quality—load-for-load. *Salted one*, with about a buſhel (½ cwt.) of ſalt to a load of hay.—The expence about 4 s. a load. *The other unſalted.*	(In Home Stack-yard.) SALTING HAY. Is Salt beneficial to Clover-hay which has been damaged in making? *Dubious.*	The Hay being put into ſtack quite dry, and the ſap being exhauſted by the weather, the ſalt did not aſſimulate with the Hay, but remained ſeveral weeks (near the outſides at leaſt) undiſſolved: and it is remarkable, that the ſtack which was ſalted did not take ſo kind a heat, nor ſettle ſo much as that which was not ſalted *
—— June and July '77. ——	—— No. LI. ——	—— May '78. ——
The ſpring being uncommonly wet, the ſoil was reduced to a ſtate of mortar. By way of experiment, however, I planted three beds whilſt in that ſtate. The reſt were planted about a month afterwards, when the ſoil was become firm.	(In S. 1.) PLANTING CABBAGES. May Cabbages be ſafely planted when the ſoil is very wet? *No: wait until the ſoil be in ſeaſon; though the time of the year may be late.*	The three beds planted in a puddle were never worth cutting: many of them went off entirely, and the reſt never throve. Thoſe planted a month afterwards overtook them before the laſt-planted had been a month in the ground: yet theſe, being planted late in July, never came to their full ſize.
——— July '77. ———	—— No. LII. ——	——— May '78. ——
a, Drumhead *American.* *b*, Red *Scotch.* Part of each ſort was tranſplanted, and part planted immediately from the ſeed-bed.	(In S. 1.) SPECIES OF CABBAGES. Are the *American* or the *red Scotch* fitter for a farm-yard? *The American,* in *Winter.* *The Scotch,* in *Spring.*	The *American* much larger than the *Scotch*; but the *Scotch* ſtood longer in the Spring. The plant proceſs was not rigidly attended to: there was no obvious difference.

* The carters *complained* that the ſalt made their horſes thirſty.

C

EXPERIMENTS OF AGRICULTURE.

THE PROCESS.	THE INTENTION.	THE RESULT.
—— 31. July '77. ——	—— No. LIII. ——	—— 4. Aug. '77. ——
Shook * the whole field into cocklits, while quite wet; except three fwaths. Made part of the cocklits fmall (half bufhels)—Part large (two bufhels).	(In T. 4.) H A Y I N G. Should Meadow-grafs which is nearly *made* and turning yellow, be fhook into cocklits *wet*; or fhould it remain in fwath? *Shook into cocklits.*	The cocklits are incomparably the beft; the fwaths are quite black. The half-bufhels dried fafteft; but the weather was tolerably fine: had much rain fallen, *perhaps* the large ones would have fared beft.
—— 1. Aug. '77. ——	—— No. LIV. ——	—— 4. Aug. '77. ——
Made the whole field into *Bufhel-cock*; except a few fwaths, which were part of them *turned*, and part left *unmoved*.	(In R. 3.) H A Y I N G. Should Mead-grafs which is fpoiling in fwath, be *cocked* or *turned*, or left *untouched?* *It fhould be cocked.*	Altho' there has not been much rain (gloomy, with fhowers; except yefterday, which was Sunday), the *cocklits* are the drieft; the *turned*, next; the *unmoved*, very bad. *The ground is damp.*
—— July '77. ——	—— No. LV. ——	—— June '78. ——
Re-fowed the whole field; except a belt acrofs the middle. (See Experiment N°. XLIV.)	(In B. 3.) RE-SOWING CLOVER. If the firft fowing mifs, is it of any fervice to re-fow when the ☉ats are in haw? *No: if the foil be out of heart and tilth.*	This latter fprinkling has been of no perceptible fervice. But this field is neither in tilth nor in heart.

* Rolling the fwaths into hard lumps, is not the proper way of making them into cocklits: They fhould be *fhook, broken,* and *mixed* with a prong, leaving the cocklits as light as poffible, with round fmooth tops, and fmall fnug bottoms.

THE

EXPERIMENTS OF AGRICULTURE.

THE PROCESS.	THE INTENTION.	THE RESULT.
—— Aug. '77. ——	• No. LVI. ——	—— Dec. '78. ——
a, a, Mowed. b, Left to be trampled by the cattle. This Experiment is repeated in three separate fields.	(In A. 2. A. 3. and C. 1.) L E Y I N G. If a Ley cannot (for want of a fence, &c.) be conveniently paſtured the firſt ſummer, ſhould it be mown; or, if the crop be *thin*, ſhould it ſtand till after harveſt to be eaten and trodden down? *Undeciſive.*	The unmown belts which were trodden down are none of them diſtinguiſhable from the parts which were mown, and the Hay, feeds, &c. carried off; except in A. 3. and even there the ſuperiority is dubious.
—— Aug. '77. ——	—— No. LVII. ——	—— June '78. ——
It rained pretty ſmartly for ten minutes, while the pullings of the Mix-graſs-ſtack were laid on: and this wet layer (1¼ load below the eaves) was immediately covered with a load which had been wetted on the waggon by the ſame ſhower.	(In Home Stack-yard.) STACKING HAY Does ſtacking Hay during a ſhower make a mouldy ſeam in the ſtack? *No.*	The whole ſtack was totally free from *mould*. And whether this ſeam was *muſty* was not properly attended to.
—— Aug. '77. ——	—— No. LVIII. ——	—— Feb. '78. ——
Stacked one layer of Hay, which had been out a month or five weeks of rainy weather, between two layers of Mix-graſs Hay, which had been well got together.	(In Home Stack-yard.) H A Y. Can bad Hay be improved by ſtacking it with good? *Dubious.*	The cattle eat the bad layer; but not ſo greedily as they do the good. I apprehend the bad received ſome benefit; but *perhaps* the good received an equivalent detriment.

THE

EXPERIMENTS OF AGRICULTURE.

THE PROCESS.	THE INTENTION.	THE RESULT.
—— 2. Sept. '77. ——	—— No. LIX. ——	—— 3. Sept. '77, ——
This evening put G. 2. into fmall cock. Left G. 1. in fwath. There is an appearance of fhowers; but not of much rain.	(In G. 1. and 2.) HARVESTING OATS. If Oats are fit for carrying, and the teams are otherwife employed, fhould the Oats be put into fmall cocks? or fhould they be left in fwath? *They fhould be left in fwath.*	There was a heavy fquall in the night; but the day has been very fine. The fwaths of G. 1. were turned about 11 o'clock, and were ready to carry by 12. The cocks of G. 2. were lightened up with a fork (the rain did not go through them) about 10 o'clock; but were not fit to take up until between 2 and 3.
—— 6. Sept. '77. ——	—— No. LX. ——	—— 6. Sept. '77. ——
A fine, hot, drying morning; but a very heavy dew. Broke a few rows of the fecond crop of Clover of Barn-Field into thick beds, before breakfaft, when the grafs was quite wet. While the labourers eat their breakfafts the dew went off, when the remainder were broken into fimilar beds.	(In I. 2.) HAYING. If in the morning the wind get up, and the fun break out hot, fhould Hay which is in cocklits be broken into beds while the dew is on the grafs? Or fhould it remain in cocklits (notwithftanding the finenefs of the morning) until the dew go off? *Wait until the grafs be dry.*	The *furfaces* of the firft-broken were prefently dry; but when we came to turn them, their *bottoms* were as wet as thatch. The laft-broken were drier in *three* hours by *one* turning, than the others were in *five* hours by *two* turnings: for the grafs being fkreened from the fun and wind, it ftill retained its moifture, and ftill kept damping the under parts of the beds.
—— Oct. '77. ——	—— No. LXI. ——	—— Aug. '78. ——
The whole field fown dry; except (*a*) brined and limed.	(In I. 1.) SOWING WHEAT. Is brining the feed of any ufe? *No.*	The whole field equally free from fmut; and without any difference in the crop.

EXPERIMENTS OF AGRICULTURE.

THE PROCESS.	THE INTENTION.	THE RESULT,
—— 12. June '78. ——	—— No. LXII. ——	—— Aug. '78. ——
a, a, Light-rolled. *b, b,* Heavy-rolled. *c, c, c,* Not rolled. The Barley is beginning to fpindle, and the roller feemed to bruife it very much ; breaking it down flat to the ground.—The corn was dry, but thee arth is moift.	(In F. 1.) ROLLING BARLEY. Should a light moory loam be rolled, when the Barley is fix or eight inches high ? *Yes : if it cannot be rolled earlier.*	Where the roller was turned (tho' a double one) it cut up many of the plants, and killed fome by merely paffing over them. The Barley, however, is now nearly of equal goodnefs ; and the parts rolled are much better to mow, and ftill better to rake.
—— 12 June '78. ——	—— No. LXIII. ——	—— Avg. '78. ——
(See laft Experiment.) *a, a,* Heavy-rolled. *b, b,* Not rolled.	(In F. 2.) ROLLING BARLEY. Does the roller injure the plants, or does it not ? *It does ; in fome degree.*	Many plants, no doubt, fuffered ; and had thefe two fields been rolled a week earlier, the crops, perhaps, would have been fomething better : as it is, they are very good.
—— 12. June '78. ——	—— No. LXIV. ——	—— Aug. '78. ——
Sowed 5 lb. of Rib-grafs, 5 lb. of Trefoil, and 3½ lb. of White Clover an acre. We had a very heavy fhower the 9th, and the ground is ftill moift.	(In K. 4.) S O W I N G L E Y - G R A S S E S. Is it, or is it not too late to fow Mix-grafs in the middle of June over Oats which have been fown eight weeks, and which are fix inches high ; the foil having been summer-fallowed and dunged for Oats and Ley-graffes ? *It is, in fome degree, but not wholly, too late.*	There is a fufficiency of plants ; but they are weak ; though apparently healthy. Dec. '78. Although we have not yet had fcarcely any froft, the plants feem to have diminifhed in number. There are, however, ftill a fufficiency to form a good meadow, provided they be paftured the firft two or three years.

THE

EXPERIMENTS OF AGRICULTURE.

THE PROCESS.	THE INTENTION.	THE RESULT.
—— June '78. ——	—— No. LXV. ——	—— Dec. '78. ——
(See laſt Experiment.)	(In K. 4.)	At preſent no determinate judgment can be formed: There is, however, no *obvious* difference between, *a*, *b*, and *c*.
a, Sown before harrowing.	SOWING LEY-GRASSES.	
b, Sown after harrowing.		
c, Not harrowed.	Should a ſtale ſurface be harrowed *before* or *after* ſowing the ſeed ?	
	Undeciſive.	
—— June '78. ——	—— No. LXVI. ——	—— Aug. '78. ——
(See laſt Experiment.)	(In K. 4.)	There is no perceptible difference between the parts harrowed, and the belt which was left unharrowed.
The harrows did not eradicate many plants; but wholly buried ſome, and partially buried many with clods and mould.	HARROWING OATS.	
	Does harrowing them when ſix or ſeven inches high, injure the plants much ?	
	No.	
—— 10. May '78. ——	—— No. LXVII. ——	—— Oct. '78. ——
a, was croſs-harrowed (by miſtake) before the croſs-plowing.	(In G. 2.)	*b*, is perceptibly the beſt fallow. The harrowing, therefore, was not only thrown away, but was applied to an injurious purpoſe.
b, Croſs-plowed, with the ridglits entire.	FALLOWING.	
a, Broke up more fallowy than *b*, which came up in large cubical clods.	Should a foul Quondal, which has been *trenched* (laid up in ridglits) in winter, be croſſed in May with or without a previous harrowing ?	
	Without a previous harrowing.	

END OF THE EXPERIMENTS.

OBSERVATIONS

CONCERNING

AGRICULTURE.

GENERAL

OBSERVATIONS

CONCERNING

SCIENTIFIC AGRICULTURE.

EXPERIENCE is CERTAIN KNOWLEDGE acquired by OBSERVATION.

OBSERVATION is the action of the SENSES upon OBJECTS, and of REASON upon IDEAS.

IDEAS may prefent themfelves fortuitoufly, or they may be intentionally fought-for :—if they prefent themfelves fortuitoufly, they may be faid to come by INCIDENT ;—if they be fought-for intentionally, they come by EXPERIMENT.

INCIDENTS and EXPERIMENTS are, therefore, the fame; except that Incidents furnifh unlooked-for information ; Experiments, that which was fought-after : but whether the EXPERIENCE be *incidental* or *experimental*, it is indifcriminately conveyed to the underftanding by OBSERVATION *.

---

* For *making an experiment* conveys no EXPERIENCE to the mind, until its *refult be afcertained* by OBSERVATION ;—any more than does one of thofe numerous *incidents* which *happen* every day, without being *obferved*.

OBSERVATION

OBSERVATION, therefore, is the procefs whereby the SENSES act upon OBJECTS, and REASON operates upon ideas ; whether the objects and ideas occur by INCIDENT, or are developed by EXPERI-MENT.—And it is allied to EXPERIENCE, as SCULPTURE is to STATUE : Obfervation is the PROCESS ; Experience, the PRODUCTION *.

OBSERVATION may act voluntarily, or involuntarily : thus, the *Philofopher* becomes wife, through DESIGN ; the *Mechanic* expert, thro' HABIT : but they are both of them led by OBSERVATION ;—the one, advertently ; the other, without knowing it.

'UNINTENTIONAL OBSERVATION may therefore give ART: but INTENTIONAL OBSERVATION can alone produce SCIENCE.

ART without SCIENCE is dependent on the MEMORY, and refts folely with ARTISTS ; SCIENCE perpetuates the ART, and transfers it, not only to diftant NATIONS, but to future AGES.

To apply this reafoning to AGRICULTURE :— the ILLITERATE FARMER either acts wholly by CUSTOM ; or, if he *obferves advertently,* trufts his obfervations to his MEMORY. The SCIENTIFIC FARMER, on the contrary, not only *obferves* and *records* the ufeful information which occurs to him in the courfe of his practice, by INCIDENT ; but *difcovers* by EXPERIMENT, thofe valuable facts, which never did, nor ever might have come, *incidentally,* within his knowledge. Nor does he reft fatisfied here ; for, barely to *record* ufeful obfervations is only one ftep fuperior to leaving them confufedly in the memory : fome METHOD is neceffary : they muft be digefted,—*fyftemized* ; or, fhould they become voluminous, they will remain, like the mifcella-neous entries of an ACCOUNTANT's JOURNAL, an almoft ufelefs chaos of valuable information. They muft therefore, like the MINUTES OF A MERCHANT, be reduced to a LEDGER-FORM—to a SYSTEM.

---

* The *mental production* is here meant : the *written* or *typographical production* is, by a familiar acceptance of the word, called an *obfervation.*

SYSTEM,

System, however, as defined by a *great* Lexicographer, " *any* com-plexure or combination of things acting together"——" *a* (or *any*) fcheme which unites many things in order," is only a vague fome-thing ; and does not by any means convey the idea here meant by A System; which is not ANY complexure, nor A fcheme; but a CER-TAIN complexure, and THE fcheme, or PLAN, WHERELY THE ELEMENTS AND PROCESSES OF AN ART ARE ARRANGED IN THE ORDER OF NATURE. Nature has only *one* PLAN; confequently, there can be only *one* SYSTEM of *one* ART.

The plan and operations of Nature, however, are by no means *obvi-ous* to human perception : They muft firft be deliberately fcrutinized by ANALYSIS, and afterwards ftrictly demonftrated by EXPERIENCE : by which alone the *analytic* OUTLINES can be improved to a *fcientific* SYSTEM. Therefore,

To TREAT THE ART OF AGRICULTURE SCIENTIFICALLY —— INCIDENTS muft be attended to,—EXPERIMENTS made and regiftered,— OBSERVATIONS performed, —INFERENCES drawn, —EXPERIENCE *recorded*,—the ART *analyzed*,—and the SCIENCE *fyftematifed.*

To *record* experience gained by INCIDENTS :——MISCELLANEOUS MINUTES, made immediately on obfervation, form a fimple RECORD OF INCIDENTS.

To *record* fcientific knowledge obtained by EXPERIMENTS: — Regiftering the proceffes chronologically, immediately on performing ; and the refults facing their refpective proceffes, im-mediately on obfervation, give a fimple REGISTER OF EXPERIMENTS.

To *analyze* an art :—Its elements and proceffes muft be feverally inveftigated, and their natural affinities traced. And from the refult of this ANALYSIS may be delineated a THEORETIC SYSTEM.

To *fyftemife* the fcientific matter recorded :—TO this THEORETIC SYSTEM,—to this LEDGER muft be POSTED such incidents and expe-riments as have been duly obferved and regiftered. By this ope-

ration,

ration, the ANALYTIC DELINEATION will not only be enriched, but *corrected* and probably *enlarged*: for as each Incident and each Experiment will require to be claffed under fome certain head or divifion (perhaps under various heads or divifions), if fuch head or divifion has not been hit upon, or has been erroneoufly arranged, by THEORY, it muft now be added, or tranfplaced; and thus on EXPERIENCE will be raifed a SCIENTIFIC SYSTEM.

By a plan fimilar to this, the Author has endeavoured to acquire AGRICULTURAL KNOWLEDGE.

During three years, he *recorded*, IN MISCELLANEOUS MINUTES, the INCIDENTS which occurred to him; and made fome EXPERIMENTS.

Affifted by a general idea of agriculture, principally received by unintentional obfervation, he had previoufly *analyzed* its elements and proceffes, and had delineated from this ANALYSIS, a THEORETIC SYSTEM.

By the means of this OUTLINE he *fyftemized* the INCIDENTS, EXPERIMENTS, and THEORETICAL OBSERVATIONS, which he had minuted. This was an operation which required both method and attention. To have tranfcribed the Minutes *verbally*, would have rendered the DIGEST almoft as unwieldy as the Mifcellaneous Minutes themfelves:— To have inferted a mere *reference*, would have made it very little fuperior to a table of contents:—A middle way was therefore adopted.— Each diftinct Minute, or each diftinct paffage of a Minute, was endeavoured to be compreffed into a MAXIMICAL SENTENCE; conveying, at leaft, the *characteriftic*, if not the *fpirit* and *intention* of the Minute referred to. As he went on with this operation, he *corrected* and *enlarged* his ANALYTIC DELINEATION; till at length he flattered himfelf, that he had FOUNDED ON TRUTH, and ARRANGED IN THE ORDER OF NATURE, fome of the PRINCIPAL COMPARTMENTS of the SCIENTIFIC SYSTEM; which compartments, more or lefs, furnifhed with the EXPERIENCE he had acquired, he ventured to fubmit to the PUBLIC.

The

The efpecial attention paid (or which ought to be paid) to a firft (and indeed to every) publication, produces an exertion of the intellects, which otherwife, perhaps, would not prevail. This, added to the light which *method* throws on the human underftanding, pointed out to the Writer, an *improvement* in the mode of acquiring agricultural knowledge. He perceived (particularly while fyftemizing the proceffes of Wheat*), that by taking a SYSTEMATIC REVIEW of the elements and proceffes appertaining to any particular production, many ideas ftarted forth, which otherwife had not prefented themfelves.

Since JULY 1777, the time of clofing the MINUTES, the Author has therefore adopted this IMPROVED MODE OF OBSERVATION: not, however, to the total exclufion of MISCELLANEOUS MINUTES on the interefting incidents which occurred to him in the courfe of his practice. —His method has been this:

If an idea ftruck him forcibly; efpecially if it was fuch a one as could not fall obvioufly within the ANNUAL OBSERVATIONS, he retained it by a rough, concife memorandum.

As the crops, whether of hay or corn, were *harvefted*, he remarked minutely their feveral degrees of *quantity* and *quality*; and, in hours of leifure, reverted to the ELEMENTS, and the PROCESSES which produced them.

Firft, he recognized the SOIL or foils they grew in: if various, he compared the crops with the fpecies which produced them, and drew his inferences accordingly. He then proceeded to the MANURE with which the foil had been fertilized: the SEED from which the crops had been raifed: and the WEATHER by which they had been influenced.

Having taken a view of the ELEMENTS, or things which were managed; he next took a retrofpect of the PROCESSES, or operations made ufe of in their management. Firft, he recalled to his memory the

* DIGEST OF MINUTES OF AGRICULTURE, page 86.

general

general management, or Succession of one crop to another, which had taken place for two or three years back : The Soil-Process, or the tillage which the foil had received : The Seed-Process, or the mode, &c. of sowing : The vegetizing Process, or the weedings, hoeings, &c. which had been given : And laftly, the vegetable Process, or the manner of harvefting, ftacking, &c.

As he took this retrofpect of the various ftages of his laft year's management, he not only minuted fuch Inferences (whether *practical leffcns*, or *theoretic hints*) as refulted immediately from the operation ; but mingled with them the Incidents which had been minuted chronologically ; as alfo the Experiments which had been made relative to the refpective crops. This laft (namely, claffing the experiments with the incidents and inferences) was a fortunate idea, which never occurred to him until he was regiftering his annual obfervations. He had fpent fome time in confidering how he fhould methodize the *mifcellaneous experiments*, fo as to render them of ordinary ufe ; not being then aware, that Incidents and Experiments are fo nearly fimilar.

On the matureft deliberation, the Author is fo fully fatisfied with this fyftematic manner of acquiring agricultural knowledge, he cannot forego acknowledging a wifh, that it may be attentively practifed by others *.

There are few men, perhaps, who have leifure and perfeverance enough to minute every ufeful incident which may occur to them *throughout the year* ; but the man who has not leifure and induftry
enough

---

* Nay, he will venture one ftep farther ; and, as every man has fome favourite reformation to enforce ; fo the author apprehends, that had he abfolute rule, one of the firft edicts he fhould iffue, would be, that every profeffor of agriculture, who did not voluntarily add to the welfare of the State, by offering up, annually, fcientific vouchers of experience drawn from his own management, fhould be compelled to promote its weal, by paying double taxes.

enough to pay unremitted attention to his farm, during *hay-time* and *harveft*, and to make ANNUAL OBSERVATIONS on his management, is by no means a man fit for a *Farmer*.

Taking for granted that there are many men who have both opportunity and inclination to increafe their own ftock, and to add to the public ftore, of agricultural knowledge; to thofe the Author addreffes the following hints.

Attending to incidents, and to the refults of experiments; and, on hours of leifure, digefting the obfervations made thereon; are, to a man wholly detached from other amufements, agreeable relaxations from the more active employments of the day : efpecially as a nicety of language is not in the leaft degree neceffary to the rough fketch to be ftruck out in harveft; for it is not the manner of relating the facts, but the facts themfelves, which are then to be particularly attended to. Befides, by letting them lie in the-rough until after autumnal feed-time, and then making out a fair tranfcript, many errors, fuperfluities, and defects, may be difcovered, which might (even on a revifion) at harveft have been overlooked : the facts will not only be now feen in a new light; but the tranfcription will root them anew in the memory. Every man who has accuftomed himfelf to *write* his ideas muft have obferved, that after his autograph has lain by him until its contents have been in fome meafure forgotten, he has, on REVISION, feen it in a new point of view : he has reviewed it, in fome degree, as the production of another; confequently with a lefs partial eye than that with which he faw it at the time of writing; and his judgment has of courfe been proportionally ftrengthened. And he muft alfo have remarked, that TRANSCRIPTION feldom fails of producing an improvement of the original.

At the time of making out the fair tranfcript, a GENERAL REVIEW of the elements and proceffes fhould be taken. This will not only call to the memory many incidents which otherwife would have efcaped

it;

it ; but will at the fame time give an opportunity of fyftemizing the incidents and experiments which have been minuted and regiftered in the courfe of the year, and which appertain to heads not peculiarly noticeable *at harveft*.

Thus a complete fyftematic view of PAST MANAGEMENT will be taken ; and a valuable collection of ufeful leffons retained, as guides to FUTURE MANAGEMENT : not, however, the theoretic dogmas, nor even the fcientific obfervations, of other men, made on other foils; but maxims drawn from felf-management, on the identical foil to be hereafter managed.

This mode of acquiring agricultural knowledge is not new : it has been more or-lefs practifed ever fince Mankind were diffatiffied with the fpontaneous productions of Nature : for it is in this manner the moft illiterate Farmer becomes knowing in his profeffion :—perhaps, however, without being aware of his acquifitions ; or, if apprifed, without providing any other means of prefervation, than merely trufting them in the care of his memory.

Every Farmer who is one link fuperior to his beafts of labour, muft fay to himfelf at harveft, " I have this year got fuch and fuch " crops ;—what has been the management ?"

As he buftles acrofs his fields, or keeps-fentry over his work-people, it is fcarcely poffible for him to refrain from reflections like thefe :

" Great Clayey Clofe is a brave piece of *Wots*: the fwaths lie rare- " and-round, i'faith ! Zuckers, and what an even plant of Clover ; and " how clean !—Ha ! there's nought beats a fummer-fallow for Great " Clayey Clofe. You may talk of ftealing a crop, indeed ; but, i'cod, " it's like ftealing from your neighbour : there's noa good comes on't " at laft."

From his lower, his bufinefs calls him to his upper farm ; where ideas like the following muft neceffarily hobble acrofs his intellects.

" The

" The Barley of Upland-Down is a brave piece of barley, that it is.
" Aye, aye, it was a rare fallow. And then the hoeings! and the
" fold!—Odds my life! a turnip-fallow and the fold againſt the
" world for Upland-Down."

If theſe incidents are ſo fortunate as to make impreſſions on his
memory, he next year manages accordingly: he *ſummer-fallows*
his *ſtiff* land, and *turnips* his *light*. He does not, however, hug
himſelf more on having made the diſcoveries, than on his being
cunning enough to keep them as family-ſecrets: and conſequently,
on being able to monopolize to himſelf and his heirs, the advantages which may accrue. The next market-day, however, may be
lucky enough to liberalize his notions; and, over the tankard, he
may communicate his ſentiments to his pot-companions; who, probably, are either too much wedded to the cuſtoms of their forefathers,
or have too high an opinion of their own management, or perhaps of
the management of ſome favourite neighbour *, to profit by the information. Thus the diſcovery either dies neglected;— remains with
the family of the diſcoverer;—or, at beſt, is introduced to a few neighbouring Farmers. Whereas, the information of the man who is con-
ſtantly on the watch for incidents who is repeatedly making experiments, and who annually reduces his experience to ſyſtem, is
not only more ample and intereſting, but he communicates his improvements to thouſands he never heard of, and perhaps to tens of
thouſands yet unborn. For it is ſcarcely poſſible for a man who
ſcrutinizes his DAILY EXPERIENCE with a SCIENTIFIC EYE, not to
make ſome uſeful diſcovery. And there is ſcarcely any man of
common underſtanding, who has carefully attended to the reſults of
his preſent management, in order to regulate the proceſſes of his
future; and who has chronologically memorized, and annually
regiſtered, theſe reſults, ſyſtematically; who muſt not in a few years
have produced a work of PUBLIC UTILITY.

* For almoſt every pariſh in the kingdom contains " the beſt farmer in England."

E   The

The Writer, however, does not prefume that each line of the following Obfervations, *taken feparately*, is of PUBLIC MOMENT ; but, *taken collectively as a model* for thofe to work by, who may not have beftowed on the *fcience* of Agriculture fo much attention as he has done, he believes there is not any thing fuperfluous : if there is, it proceeds from an error of judgment, not of defign.   He does not, however, hold out the following, as a model perfect beyond the power of human invention to improve (he has repeatedly *altered* it himfelf, and it may perhaps hereafter be repeatedly *altered* by others) ; but he ventures to predict, that it will not readily receive any *effential improvement*.

This prediction, however, is not thrown out fo much to gratify the ambition of the Writer, as to obtain a *fecond Perufal* from the Reader, who may not at firft fight, fee the utility of this mode of acquiring agricultural knowledge :   and more efpecially, with an earneft defire of drawing down the attention of thofe who may be too *highly* read in the prefent *ton* * of Literature to drudge, unfolicited, at inveftigation.

It feems neceffary to remark in this place, that the following OBSERVATIONS, notwithftanding the difference in form, are, in reality, a *Continuation* and *Conclufion* of the MINUTES OF AGRICULTURE, and may be confidered as *fupplemental* to that Publication.   The Author, however, has endeavoured as much as poffible to render this *a feparate Work*, in order to accommodate the Reader who may not be poffeffed of the MINUTES; the *References* to which, being merely intended to *explain* or *corroborate*, are rather convenient than neceffary to this Publication.

* The Reader has free permiffion to make ufe of any other Gallically jargonical cant of the day he pleafes : The Writer means the prefent prevailing fafhion of *Letter-reading* which characterizes this novel-fipping, tour-tippling, voyage-devouring period.

OBSERVATIONS

CONCERNING

AGRICULTURE.

I N AUTUMN,—prior to Wheat Seed-time, the Writer has made it a Rule to sketch out the Plan of his next Year's Management, by delineating an INTENDED ARRANGEMENT. This *theoretic* Plan, however, he has never considered as perfect and *inviolable*; but has continued *altering* and *improving* it, *as Circumstances pointed out in the Course of his Management.* He, nevertheless, has always found it of very great service in proportioning his work to his teams;—the number of acres to be plowed, to the number of beasts of labour he has had to plow them with: besides having a more distinct view of the business of the coming year, than he could have had without such a Sketch. The utility of this *intended Arrangement* will appear more fully when the REAL ARRANGEMENT and its Uses are pointed out.

IN SEED-TIME,—the following has been his constant practice: As soon as the sowing of any particular field is finished, he first adjusts and closes the *Labour Account* of that field; (See DIGEST, page 145.) and, having previously opened a *Seed Account* * for each of the Crops intended to be sown next year, he registers in one line (as in the following Arrangement) the Time of Sowing, the Number of Acres, and Name of the Field; with the Quantity and Quality of Seed which has been sown in it. As soon as the whole of a crop, as

---

* The References to these *Seed Accounts* were omitted (by a typographical error) in the *Index* which was given in page 145 of the DIGEST.

Wheat,

Wheat, for inftance, is fown, he adds up the quantity of acres and the quantity of feed fown over them; and thus fixes the real Arrangement with refpect to *Wheat.*

These feveral Operations, and this Arrangement, fet the Soil and Seed Proceffes in a clear and interefting point of view; much ufeful information neceffarily arifes; and many incidents now require to be retained, until Harveft, by rough Memorandums.

In MAY,—or as foon as the Seed is all in, he takes a *general View of the whole Farm*; correcting fuch departments of the intended arrangement as have not fallen under the *Seed-Procefs*;—as *Meadow, Pafture, Fallow,* &c. and thus afcertains, precifely, the REAL ARRANGEMENT.

THE

ARRANGEMENT,

1777.

WHEAT.

Time of Sowing.		Quantity of Seed.
29—9 to 30—10*.	24 Acres in L 1, 2.	49½ Bufhels of M†.
24—9 to 7—11.	18½ ———- P 1, 2, 3.	41¼ ——— B.
28—10.	1 —— of S 2.	2 ——— M.
	43½ Acres	93 Bufhels of Seed.

* The time of fowing L 1 and 2 was from the 29th of the *Ninth* month (September) to the 30th of the *Tenth* month (October) 1776.

† Of Wheat which grew in the divifion M.

WINTER-

WINTER TARES.

Time of Sowing.		Quantity of Seed.
October 10.	3 Acres in O 1, 3.	7 Bushels of G.
30—10 to 2—11.	1½ ———- S. 1.	3 ——————
	4½ Acres	10 Bushels of Seed.

BARLEY.

November 8.	o½ Acre of S 2.	1 Bushel of O.

PEA-BEANS.

25—2 to 1—3.	10¼ Acres of M. 3, 4, 5, 6.	43 Bush. of Maz. Beans of L ; and Marlborough Peafe : half-and-half.

TARE-BARLEY.

Mar 6. to 22.	8¼ Acres in M 1, 2.	26 Bush. ⎰ Saved from laſt
27.	5¼ ———- F 2.	18 —— ⎱ year's Fodder of L.
	13½ Acres.	44 Bushels of Seed.

OATS.

24—3 to 5—4.	·8¾ Acres in B 3 and 4	⎰ 15 Bush. off a Chalk. ⎱ 2½ ———- raiſed in E. ⎱ 28 ——- Alemouth.
29—3 to 6—4.	5½ ———- A 4 and 5.	26 ——- Alemouth.
22—3 to 17 4.	10 ———- G 1 and 2.	⎰ 25 ——- Chalk. ⎱ 32 ——- E.
	24¼ Acres.	128½ Bushels of Seed.

POTATOES.

15—4 to 1—5.	1¼ Acre in S 3.	⎰ 10 Bush. of Red Bunch. ⎱ 11 ——— White Kidney
	1½ Acre.	21 Bushels of Plants.

CLOVER.

ARRANGEMENT, 1777.

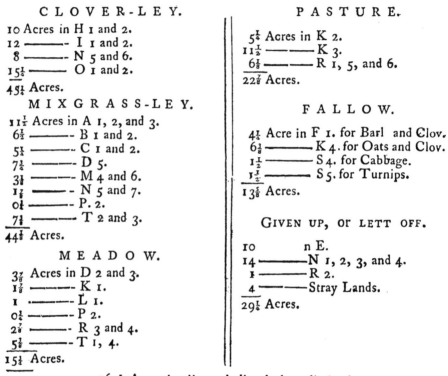

CLOVER-LEY.

10 Acres in H 1 and 2.
12 ———- I 1 and 2.
8 ———- N 5 and 6.
15¼ ——— O 1 and 2.
45¼ Acres.

MIXGRASS-LEY.

11½ Acres in A 1, 2, and 3.
6½ ———— B 1 and 2.
5¼ ——— C 1 and 2.
7¼ ——— D 5.
3¼ ——— M 4 and 6.
1⅞ —— N 5 and 7.
0¼ ——— P. 2.
7¼ ——— T 2 and 3.
44⅞ Acres.

MEADOW.

3⅞ Acres in D 2 and 3.
1⅞ ——— K 1.
1 ——— L 1.
0¼ ——— P 2.
2⅞ ——— R 3 and 4.
5⅞ ——— T 1, 4.
15¼ Acres.

PASTURE.

5¼ Acres in K 2.
11½ ——— K 3.
6⅛ ——— R 1, 5, and 6.
22⅞ Acres.

FALLOW.

4¼ Acre in F 1. for Barl and Clov.
6⅛ ——— K 4. for Oats and Clov.
1½ ——— S 4. for Cabbage.
1½ ——— S 5. for Turnips.
13⅝ Acres.

GIVEN UP, or LETT OFF.

10　　　 n E.
14 ———N 1, 2, 3, and 4.
1 ———R 2.
4 ———Stray Lands.
29¼ Acres.

269⅞ Acres in all, excluding hedge, ditch, &c.
21 ——— of hedge, ditch, roads, &c.
291 ——— including waſte.

Thus every rod of the Farm is arranged under the head to which it immediately appertained in the Year 1777. By this mode of Arrangement, every patch and every corner is brought into view ; no part, be it ever ſo minute, can eſcape notice ; no ſtraggling acre of fallow can be left unſtirred ; no ley forgot to be rolled ; nor corn omitted to be diſweeded : the eye, at one glance, takes in the whole economy of that year's management.

In

In Harvest,---he opens an *Account*, or *Head*, for every particular crop, or vegetable, to be harvested. And as the hay or corn is *carried*, he regifters, in the evening, the number of loads which have been carried during the courfe of the day; mentioning in one line, the month and day, the field, the number of *Harveft*-loads; and, of *Hay*, the eftimated number of *Sale*-loads; and *gueffes* at the number of Quarters the *Corn* crops will yield. When the whole of any particular crop is carried, he adds up the real number of harveft jags, and the fuppofed number of fale-loads of hay, and quarters of corn; and thus afcertains the *grofs Produce* of that crop.

These *Crop-Accounts*, or *Accounts of Produce*, he either keeps mifcellaneoufly, and afterwards digefts them agreeable to the repofitories, whether barns or ftacks, to which the loads have been carried; or, which is more expeditious, he fubdivides the heads according to the ftacks he means to make, or the barns he intends to fill, with the refpective crops, and *carries* with his pen the number of loads immediately to the *Barn* or *Stack* to which the Hay or Corn had been carried by the waggons. But ufeful as *thefe* accounts of Froduce are in the *Barn* and *Farm-yard Management*, they do not give a diftinct idea of the *produce of each field*; he therefore *re-claffes* them, fo as to afcertain, precifely, the number of loads produced by each field or divifion; in order to form a comparative judgment of the various fpecies of management which have attended the different departments of the Farm; and from thence to draw Lessons of future Management.

CLOVER

CLOVER,

1777.

Mowed.			Carried.			Laid at	
June 13 and 14.	5¼ Acres in I 1.	July 15 to 31.	2½ Field Jags.	2 Sale-loads
16 to 24.	6½ —— I 2.	—— 11 to 15.	14 ——	9¼ ——
16 to 24.	10 —— H.	—— 11 to 15.	20 ——	14½ ——
25 to 30.	15¼ —— of O.	—— 14 to 16.	21 ——	14¼ ——
July 14 and 15.	8 —— N.	—— 18 and 19.	11 ——	10½ ——
	45¾ Acres.		68½ Jags.	51¼ Loads.

SOIL.

Part Loam (I);

Part fandy Loam (H);

Part gravelly Loam (O, N).

I find, on comparifon, that the Crops are nearly in proportion to the tilth and heart of the refpective Soils; the ftrongeft Loam, however, feems to have the preference.

This year's crop of Clover, therefore, affords no *pofitive*,---no *certain*,---no *fcientific inference*, with refpect to *Soil*: But, from the refult of this year's experience, and from a feries of fimilar obfervations, I will venture to memorize the following *probable*,---*prefumptive*,--*theoretical inferences*, as *hints* which may lead to more certain knowledge in the courfe of future management.

Perhaps ;--- *Clover affects almoft every fpecies of Loam.* But,

Perhaps ;---*The ftronger the Loam, the better it is affected by Clover.*

* The Dates of Mowing and Carrying are here particularized, to fhew the Tedioufnefs of the Clover-Hay-Time 1777. But as they render *the Produce of each Field* lefs perfpicuous, they are hereafter omitted.

4

MANURE.

MANURE.

Part was dunged for the prefent crop ;
Part undreffed.

By Experiments, No. 24 *to* 28 ;—*Compoft* laid on *the furface of a* ftrong Loam, in winter, *was of no fervice to Clover.* But,

By Incidents, in H and O ;—*Compoft* laid on *a* fandy Loam *or a* gravelly Loam,. in winter, *was very beneficial.*

SEED.

The whole raifed on a clayey Loam.

WEATHER.

Autumn, was very fine.
Winter, temperate, and remarkably dry : (See MINUTES of 19 Feb. 1777) fome fnow, but very little rain.
Spring, moift : February-March, remarkably mild ; March-April, cold.
Summer, uncommonly wet. (See MINUTES of 23 June and 15 July 1777.)
April May being dry, and the rains fetting in fuddenly, the Clover was forced haftily up in a few weeks.—This made the plants run very much to *ftem,* with little *leaf,* and ftill lefs *head,* or flower. The quantity of the crop was great ; but its quality was coarfe and fpiritlefs : And generally,

Perhaps ;—*Clover requires* moifture, *rather than* wet.

F   SUC-

CLOVER, 1777.

SUCCESSION.

Part was after wheat and a fummer-fallow, (H and I 2):
Part after barley and a Dog-day's-fallow, (N):
Part after clover, after barley and a very good winter and fpring-fallow (O):
Part after barley, and a middling winter and fpring fallow (I 1):
(H and I 2) and (N) hit very well: (that is, the furface was fur-nifhed with a fufficiency of young plants.)
(O) (two year's old) a partial crop.
(I 1) almoft totally miffed.

Therefore *,---*Clover requires a kind of pulverous tilth.*

This inference is not drawn from laft year's experience only, but likewife from incidents, this year, in B.—B 4. was fummer-fallowed; B 3. once plowed. They were fown with the fame feed on the fame day: the former has an abundance of ftrong, healthy plants; while the latter has fcarcely ten vigorous plants in the whole field;—though fpikey-rolled, and repeatedly harrowed.

SOIL-PROCESS.

See SUCCESSION.

SEED-PROCESS.

Part was fown over wheat, in April;
Part, over barley, in April and May, before the corn was up;

---

* Inferences may be divided into *pofitive* and *probable :* In the courfe of thefe Obfervations, the former are marked with *Therefore,* the latter with *Perhaps.* But even fuch as are preceded by *Therefore* muft not be received as mathematical Corol-laries, and trufted as *infallible* guides. The moft the Writer can fay in their favour is, that he means to abide by them as *true,* until he finds them, *by future experience,* to be *falfe.*

CLOVER, 1777.

Over wheat, all good ;

With barley, fome good, fome very bad.   Therefore,

Perhaps ;---*It is more certain to fow clover after the corn is up, than on the naked furface* *

By Experiment No. 1 ;—*It is not material whether Clover-feed, fown on the bare furface, be harrowed and rolled, or only rolled in.*

By Experiment No. 4 ;—*Clover may be fown, with fuccefs, over wheat, without harrowing or rolling.*

VEGETIZING PROCESS.
Rolled ; and difweeded of docks.

VEGETABLE PROCESS.

Began mowing the 13th of June, *with an appearance of wet weather* ; *becaufe* the crop was lodged, and muft have rotted on the ground ; and *becaufe* I wifhed to have the fecond crop off in time to fow it with wheat in September.   I ftopt the mowers as foon as they had cut all that I intended to fow with wheat the enfuing Autumn, but not before.---The principal part had from a fortnight to three weeks rain.

The fwaths were broken into cocklits as faft after mowing as fair weather would permit, and the bared furface raked,---wet-or dry.

---

* In this, as in many other inftances, which frequently occur in the abftrufe fcience of Agriculture, the fame effect may be produced by various caufes : Thus, the bad fuccefs of the Clover in (I 1.) (See the *Succeffion*) might be owing to its bad ftate as to *tillage* ; or it might have been caufed by a fwarm of *Infects* which happened to ravage that particular field, without molefting the others.   And nothing but a variety of fimilar refults can warrant a *pofitive Inference.*

Every

CLOVER, 1777.

Every fair blaft was embraced to turn the cocklits ; and, when the weather changed, no pains were fpared to put it ruftlingly dry into ftack*.

Finifhed carrying the 26 July.   One ftack was *falted.*

By Experiment, No. 50 ;—*Salting Clover-hay, which had been damaged by the weather, was,* apparently, *detrimental to it.*

CROP.

See the head of this fection.

QUONDAL.

(N, H, and I 2) very clean ;

(O) foul ;

(I 1) very foul.

From a GENERAL REVIEW of *this,* and of *former years Experience,* I have, while the Clover-management is now frefh in my mind, formed the following RESOLUTIONS, which I intend ftrictly to abide by, until by *future Experience* I find them to be ill-founded. They are by no means *fcientific Inferences* drawn from certain facts ; but merely fuch leading features of the Clover-management as ftrike forcibly, after having dwelt for fome time, exclufively, on that particular fubject †.

---

* It is remarkable, that while the hay in H, a low fituation, was perfectly dry, that in O, a high fituation, was quite damp : — it *made* much fafter in a low than in a high fituation. At *Woodmanfton,* a ftill higher fpot, it was at leaft one day of fine weather behind *Adfcomb,* and two behind *Croydon Common.* How is this to be accounted for ? A great quantity of rain has fince fallen.

† This is meant as a *general Explanation* of fuch TEMPORARY RULES as may, at the clofe of each article, follow the adverbial phrafe " in future."

CLOVER, 1777.

IN FUTURE,

I will not venture Clover on land which is out of tilth, or much out of heart.

I will not sow the seed until the corn be up, to defend the young plants from the Fly.

I will not manure a retentive soil, in winter, for Clover.

I will not let Clover stand too late, if it is intended to be succeeded by Wheat.

I will not suffer it to lie on the ground in swaths, except the weather be exceedingly fine.

Nor carry, until rustlingly dry ; I will rather risk its spoiling in the field than in the stack*.

---

* Some weedy rubbish, which (being referved for the top of the ftack) ftood near two months of rainy weather in the field, is ftill *Hay* ; though it ftood a month in large (three-pitch) cocks without moving.   The cock-fteads are as bald as a fallow ; and the fkirts of the cocks had all the appearance of dung ; but after it had been *thoroughly broken* and *thoroughly dried;* its appearance totally changed.   I offered fome of it to the horfes of the team which carried it, and they eat it greedily ; though *originally* very bad hay.

MEADOW.

M E A D O W,

1777.

Field	containing (Acres)	produced * (Jags)	laid at (Loads)	or truſſes an Acre.
K 1.	$1\frac{7}{8}$ Acres.	$3\frac{1}{2}$ Jags.	$3\frac{3}{4}$ Loads.	72 †
T 1, 4.	$5\frac{3}{8}$	10	$9\frac{1}{2}$	63
D 2, 3.	$3\frac{7}{8}$	6	$5\frac{3}{4}$	53
P 2.	$0\frac{3}{4}$	1	1	48
L 1.	1	$1\frac{1}{2}$	$1\frac{1}{4}$	45
R 3, 4.	$2\frac{7}{8}$	4	$3\frac{1}{4}$	40
	$15\frac{3}{4}$ Acres.	26 Field Jags	$24\frac{1}{2}$ Sale Loads.	

S o i l.

Part clayey loam (DK, L);
Part loam (T);
Part gravelly loam (P, R).

M a n u r e.

Part mudded (D 2);
The reſt undreſſed.

By Experiment, No. 21;—*Pond-dregs laid on a clayey Meadow, in November, are of no obvious ſervice.*

S e e d.

W e a t h e r.

(See Clover, page 17.) The Meadow-hay this year is more in Quantity, but of a much worſe Quality, than uſual.

Therefore;—*A very wet Summer is pernicious even to Meadows.*

* Began mowing the 18th of July, and finiſhed the 25th. Began carrying the 4th of Auguſt, and finiſhed the 8th of Auguſt.

† Very coarſe and ſedgey; this part lying very wet, and being ſubject to be overflowed.

S u c-

SUCCESSION.

Part paſtured laſt year (T).
The reſt mowed.

SOIL-PROCESS.

SEED-PROCESS.

VEGETIZING PROCESS.

Swept and rolled; and diſweeded of docks, when the ſurface was moiſt : the expence of diſweeding not one ſhilling an acre; and the benefit to this, and to ſucceeding crops is obvious.

Therefore;—*Diſweeding Meadows is a valuable operation, performed at a ſmall expence* *.

But it ſhould be done before the graſs get too high; otherwiſe, the mowers will be foiled, and the crop injured.

VEGETABLE PROCESS.

The Mead-graſs, too, (See CLOVER) was cut while the weather was unſettled ;—but the crop was full-grown, harveſt was at hand, and, from the quantity of rain which had lately fallen; I judged that the chance was in favour of fair weather : I was, however, in part miſtaken; a remnant ſtill remained, and the principal part of this hay had ten or twelve days rain.

The wet came ſo inceſſantly, there was only one day on which the graſs could, with any propriety, have been got off the ground ; and unfortunately, that day happened to be the day of indolence. This lovely hay-day being loſt, it lay in ſwath till quite yellow, and,

---

* This mode of deduction may, to ſome, ſeem unneceſſarily *formal*; but thoſe who are aware that *perſpicuity* ariſes from *method*, can readily excuſe *the formalities of ſcience.*

under

under the hedges, beginning to turn black. The firſt fair day the ſwaths were broken into buſhel-cocks: This helped it very much; and, had the cocks been turned and ſhook up next day, the benefit would have been ſtill greater; but it was jubilee-day again! and, being once ſpoilt for want of breaking out of ſwath, it was again injured for the want of being broken out of cock. The firſt day of induſtry, it was ſhook into beds, turned, and carried into ſtack ruſtlingly dry.

By Experiment, No. 53;—*Meadow-graſs, which is nearly made, ſhould be broken into cocklits, though ever ſo wet.*

By Experiment, No. 54;—*It is better to* cock *it, than either to* turn *it, or to leave it* unmowed.

But this Hay-harveſt has been *remarkably wet,* and theſe Experiments may not, *generally,* be deciſive.

CROP.

That which was paſtured laſt year, gave the largeſt quantity, (the coarſe Patches excepted) and the beſt herbage.

QUONDAL.

In the middles of ſome of the ſwath-widths, each gaſh of the ſithe appears, even this wet year, bald, dead, and unſightly.

IN FUTURE,

I will endeavour to paſture and mow Meadow-ground, alternately.

I will not manure a ſtiff-land Meadow in winter.

Embrace every favourable opportunity of diſweeding a Meadow before the graſs get too high.

See

MEADOW, 1777.

See that Meadows be cut level, *but not* fhaved too clofe.

If the ground be cold, *I will cock immediately after the fithe, or as foon as actual rain will permit.*

If the ground be hot, *and the weather fine, I will let the fwaths lie to wither,—break them into beds,—turn,—carry.* But,

Be the ground hot or cold, and the weather fettled or unfettled, I will not carry, *until ruftlingly dry.*

For although the fwaths, during the rain, looked yellow and perifh ing, when fine weather came, the hay refumed, in fome degree, its greennefs, and takes a very kind heat in ftack.

G   BARLEY,

B A R L E Y,

1777.

Sowed.   Carried.   Laid at,
Nov. 7. o¼ Acre of S 2.——Aug. 8. 1 Field-Jag,——2 Quarters:

S o i l.

A black, porous, sandy loam; with a cold, springy subsoil.

M a n u r e.

Well dunged for the preceding crop.

S e e d.

Common Barley: raised in P 3, in 1774; in T 3, in 1775; and N 7, in 1776.

W e a t h e r.

See Clover, page 17.

S u c c e s s i o n,

After Potatoes.

S o i l P r o c e s s.

The potatoes dug up, the couch, &c. picked out, and the soil gathered into half-rod ridges for the Barley.

S e e d P r o c e s s.

Sown *broadcast* the 7th of *November*.
Part was sown over the *fresh plit*; part over the *fresh plit, fluted.*

Part

BARLEY, 1777.

Part of the feed was fown *dry*; part was *pickled* and limed.

The quantities of feed, equal :—about 2⅛ bufhels an Acre.

By Experiment, No. 20 ;—*Barley may be fown* in November, *with fafety.*

For S 2. the patch on which this grew, is the wetteft fpot on the farm.

Laft year I was convinced that Barley will ftand a *very* fevere winter, in a *dry* fituation, fown in *September.* (See MINUTE of 2 FEB. 1776.) And this year I am as fully convinced that it may be fown fo late as *November*, in a very *wet* fituation. It is true, the fevere weather of laft *January* made it droop and look pale,—many of the tips being quite nipped by the froft;—but as foon as the weather became warm, it refumed its vigour as if it had not been difeafed : I apprehend, however, from the *Thinnefs* of the crop, that fome of the plants perifhed. Therefore,

Perhaps ; *Barley fown in Autumn fhould be fown thicker than that fown in Spring :*

It did not feem to *tiller* fo much as fpring-fown Barley.

By Experiment, No. 20 ; —*Barley fown over* Flutes *gave an* evenner *crop than that fown on the rough* plit.

By the fame ;—*The part fown with* dry *feed was equally as good as that fown with feed which had been* pickled.

VEGETIZING PROCESS.

None : the crop being remarkably clean, even from *ketlock*; which, had it been fown in Spring, would have vegetated abundantly. Wheat, Rye, *&c.* are always much freer from ketlock

provided

BARLEY, 1777.

(provided they are neither harrowed nor rolled in the fpring) than fpring crops are.

VEGETABLE PROCESS.

Mowed the 4th, and carried the 8th of Auguft; and was thrafhed out immediately for thatch for the meadow-hay.

CROP.

A middling field-load from rather more than half an acre.

QUONDAL.

Very foul; though the Soil was *hand-picked* after potatoes. (See MINUTE of 27. OCTOBER, 1776.)

GENERAL OBSERVATIONS.

The *Grain* is not remarkably fine. This, however, is not owing to the *time of fowing*, but to the *fpecies of foil*; for laft year an adjoining patch, fown in fpring, was not nearly fo good a fample: and the *ftraw* is this year invaluable. Many ftacks ftand naked for want of thatch, and muft remain uncovered until new wheat be thrafhed out. The ftraw from this autumn-fown crop is long, remarkably ftrong, and nearly as good as wheat-ftraw for thatching ftacks with.

Therefore; *It is convenient to have a piece of autumn-fown Barley for thatch for the meadow-hay and the corn-ftacks* *.

This year, it is true, the meadow-hay was cut late, or it could not have received the benefit of this ftraw; but the Barley was like-

---

* The Writer hopes not to miflead by this or any other inference. With refpect to fowing common Englifh Barley in Autumn, he can fay with truth, that he has tried it with fuccefs two feparate years, the winter of one of them being uncommonly fevere: but it muft be recollected, that thefe Experiments were made *to the fouthward of London*; in the more *northern parts of the Ifland*, the trial fhould be repeated with diffidence.

wife

wife cut late: had it been a forward harveft, it would have been ready a fortnight or three weeks ago. Laft year it was cut the 21ft of July; and in any year it will come foon enough to be thrafhed before the meadow-hay has done fettling; and the firft cut of clover is the only crop it will not cover.

IN FUTURE,

If leifure, the weather, and the ftate of the foil, permit, I will fow Barley in Autumn.

If the foil be yet foul, the teams well employed, or the weather unfavourable, I will defer fowing till Spring*.

---

* *Perhaps*, the practice of fowing Barley in the Spring has arifen from its *conveniency*, rather than from its *neceffity*. It being found neceffary to fow Wheat in Autumn, it became convenient to defer the fowing of Barley until Spring. The practice becoming univerfal, Cuftom took it for granted that Barley would not ftand the winter.

TARE.

T A R E - B A R L E Y*.

1777.

Five Acres and three-quarters in M. being exceedingly foul with ketlock, it was, in the beginning of July, while too young for hay, either cut for Verdage, or plowed in as an herbaceous melioration.

F 2. } containing { 5¼ Acres. } produced† { 10 Jags. } laid at { 9 Loads. } or truffes an-Acre. { 62.
M 1. } ___ { 2½ ___ * } ___ { 4 ___ } ___ { 3½ ___ } ___ { 58.

S O I L.

Part a black, porous, fandy loam, lying wet (F.).

Part a gravelly loam, lying very dry (part of M. 1.).

The reft a ftrong, fandy loam, lying moift. This part was foiled-off or plowed-in.

M A N U R E.

F 2. was *very well dunged* ‡ for Wheat of 1775,

Part of M 1. was *dunged* for Wheat of 1776.

The reft was *dunged* for Peafe of 1775.

S E E D.

* A mixture of *Tares* and *Barley* fown early in the Spring, as a fallow-crop, to be cut for *Hay*, while the Tares are in bloffom. See a variety of MINUTES referred to in page 91 of the DIGEST.

† Mowed 7th and 8th Auguft. Carried 11th and 12th Auguft.

‡ There is not any department of the fcience of Agriculture which is more indefinite than this which appertains to the *Quantity of Manure*. A *Load of Dung*, or other Manure, is almoft as vague as a *Parcel*, or a *Heap of Dung*: nor does afcertaining the exact *gauge* of the cart it is carried in render it definite; for the

quantity

TARE-BARLEY, 1777.

SEED.

Saved from the Tare-Barley Fodder of L 1. (See a MINUTE of 5 APRIL, 1777.)

Sown over the frefh plit in March.

The quantity of feed about 3¼ bufhels an acre.

WEATHER.

See CLOVER, page 17. The fecond week of July was tolerably fine; but the rains did not ceafe until the 1ft of Auguft; after which the fummer was as uncommonly fine, as it had before that time been uncommonly wet: during Auguft and September there was not, generally fpeaking, a drop of rain!

SUCCESSION.

F 2. was after Tares, after Wheat. (After *Tares*, becaufe a fallow-crop was neceffary to clafs this field with F 1. a fummer-fallow.)

M 1. after Wheat, after Peafe.

After Tares the Barley was very good, but the Tares were thin and puny.

After Wheat the Tares were luxuriant and abundant; though the feed was identically the fame.

Therefore;

quantity carried *above* the cart is ftill undetermined: not even if the height of the ridge above the cart was defcribed; for a *wide* cart will, in the ordinary manner of laying-up a dung cart, carry a much larger load than a *narrow* one of the *fame gauge*. Under the article MANURE in the GENERAL REVIEW of 1777 and 1778, this fubject will be more fully treated of: at prefent, however, it is neceffary to fay, that when a field is faid to have been "dunged," it is meant that fuch field received *about* (ten loads of dung, each load confifting of 50 cubical feet, or) 500 *cubical feet of dung*: "well dunged," means 6 or 700; "very well dunged," 7 or 800; "had a "fprinkling of dung," 4 or 500; "a fmall fprinkling of dung," 3 or 400 cubical feet of *digefted* dung, or a quantity of *undigefted* dung, or *Compoft*, equivalent thereto.

TARE-BARLEY.

Therefore; *Tare-Barley after Tares, without an interme-diate dreffing, is bad management.*

SOIL-PROCESS.

F 2. was broke up, in July, by Trenching*; in Auguft, it was crofs-plowed; in October, rough-harrowed; in November, rolled, re-harrowed and landed up; in March, thefe lands were reverfed for the feed.

M. once plowed.

SEED-PROCESS.

Part fown over the *whole* plit.

Part over the *broken* plit.

Part over *Flutes.*

By Experiments, No. 35, 36.----*It was not very material whether the plits were* broken, *or* left *whole, or* fluted, *for Tare-Barley.*

VEGETIZING PROCESS.

Difweeded of docks and thiftles.

VEGETABLE PROCESS.

Part, being uncommonly full of ketlock, was plowed-in or mown for Verdage.

Part mown for Hay.

Began to cut the 5th of Auguft †, when the Tares had juft done blowing, and the Barley was beginning to change. It was broken into cocklits immediately after the fithe,--turned, and fhook up;--carried without rain, and ftacked with the mix-grafs hay of Norwood, layer-for-layer, for fodder for the working-oxen.

* In fome Counties called *raftering.* See a Note, page 66. of the DIGEST.

† It muft be obferved, that the Harveft of 1777 was at leaft a fortnight later than ordinary. Tare-Barley, fown in March, may, in general, be cut for hay in the middle of July

CROP.

CROP.

See the Head of this Article.

QUONDALS.

Each of them fufficiently clean to be got ready for Wheat, if required; and may therefore be brought to a gardenly ftate before they will be wanted for Barley and Clover, in April or May.

IN FUTURE,

So long as I work Oxen on land which will not bear the fold, I will feed them on Tare-Oat or Tare-Barley Fodder.

Perhaps; *If the foil be clean enough to require only one plowing, fow Tare-Barley in Autumn.*

It will come fomewhat earlier than the Spring-fown, and will confequently give room for a fuller Dog-days Fallow.

H   PEABEANS

PEABEANS.

1777.

$$
\begin{matrix}
\text{M } 5. \\ 4. \\ 6. \\ 3.
\end{matrix} \Bigg\} \text{ containing } \begin{cases} 1\frac{1}{2} \text{ Acre.} \\ 2 \text{ ——} \\ 3 \text{ ——} \\ 4 \text{ ——} \\ \overline{10\frac{1}{2} \text{ Acres.}} \end{cases} \text{produced } \begin{cases} 3\frac{3}{4} \text{ Jags.} \\ 4\frac{1}{4} \text{ ——} \\ 5\frac{1}{4} \text{ ——} \\ 7 \text{ ——} \\ \overline{20 \text{ Field Jags.}} \end{cases} \text{laid at } \begin{cases} 3\frac{1}{2} \text{ Quarters.} \\ 4\frac{1}{2} \text{ ——} \\ 6\frac{1}{4} \text{ ——} \\ 8 \text{ ——} \\ \overline{22\frac{1}{2} \text{ Quarters.}} \end{cases} \text{ or bushels an Acre. } \begin{cases} 19. \\ 18. \\ 17. \\ 16. \end{cases}
$$

SOIL.

A hot Gravel, tolerably clean, interfperfed with cold, fpringy patches, fome of them very foul.

The *gravelly* parts produced the moft *Peafe*, the *loamy* parts the moft *Beans*. The Gravel, on the whole, the beft Crop,—*this very wet year*,—and by much the cleaneft Quondal.

On the *gravelly* parts, not one *Bean* in ten came to perfection :

Therefore ; *Beans, even Mazagan Beans, in a wet year, do not affect gravel.*

Therefore ; *Peabeans are unfit for a gravelly foil ;*

For the feed *Beans* are chiefly wafted.

MANURE.

M 5, was *dunged* for the Crop of 1776.

The reft *very well dunged* for the Crops of 1775.

SEED.

Two bufhels of Mazagan Beans, (raifed in L.) and two bufhels of Marlborough Gray-Peafe (immediately from Wiltfhire), an Acre; excepting one land fowed with clean Peafe, and another with entire Beans.

* Began cutting the 7th, and finifhed the 15th of Auguft. Began carrying the 14th, and finifhed the 20th of Auguft.

By

PEABEANS, 1777.

By Experiment, No. 33;—*Peafe are more eligible than either Beans or Peabeans, for a gravelly foil.*

WEATHER.

See *Tare-Barley*, page 17. Many Peafe were this year very much blighted ; fome fields went off entirely, and others were greatly injured. Perhaps this malady was owing to the chilling colds of March-April. This opinion is the more probable, as a peculiarly cold, fpringy part of M 3, was wholly cut off, while the warm gravelly parts of the fame field were not perceptibly injured. Peafe in general, which efcapéd the blight, or furvived under it until the warm rains of fummer fet in, are this year a *large* Crop ; though in general they feem to be but indifferently *podded*. Therefore,

Perhaps ; *A cold Spring is peculiarly unfavourable to Peafe.*

Perhaps ; *A wet Summer is productive of halm.* But,

Perhaps ; *A nice proportion of warmth and moifture is neceffary to the production of a* yielding *crop of Peafe.*

This laft is a very probable inference, when we reflect. how feldom we have a *Pea Year*.

SUCCESSION.

M 5, was after Wheat, after a Summer-fallow.

M 4, after Wheat, after Peafe and a Dog-days fallow.

M 3, 6, after Wheat, after an old Rye-grafs and Clover-ley, Dog-days fallowed.

The Crops of the different fields were nearly on a par. It is very remarkable, however, that M 5, which, inftead of being exhaufted by a Crop, had been dunged, in 1775, was barely fuperior to M 4,

H 2

which

PEABEANS, 1777.

which was *Peafe* in 1775, and which had received no intermediate
dreffing between the two Crops of Peafe.

As to M 5, it muft be obferved, that although this little paltry
field had received a Summer-fallow, it (being uncommonly full of
Couch, and too narrow to be crofs-plowed) was ftill foul. Befides,
it had been worn out by a fucceffion of Corn Crops, while the other
fields had been Ley : nor was the quantity of dung proportionable
to that which the other fields had received : add to this, the Manure
of M 5, had been *buried in*, while that of 3, 4, and 6, had been *laid on*
an *abforbent* foil.

SOIL-PROCESS.

To try the effect of cleanfing the furface of a CORN QUONDAL,
intended to be fown with Pulfe on one plowing, part of thefe fields
was wholly difcumbered of the ftubble, weeds, &c. part was parti-
ally cleanfed, and part left rough.

By Experiment, No. 30; — *The part wholly difcumbered
fluted the beft.*

By the fame; — *Loofening the furface rendered the foil too
fallowy for Peabeans.*

This, however, was in fome meafure owing to its being plowed
in too narrow plits.

By the fame; *Although difcumbering the furface did not add
to the* Crop, *it was of fervice to the fucceeding* Quondal..

SEED-PROCESS.

Part of the feed was fown over *Flutes*; part over the *frefh Plits.*

By Experiment, No. 32;—FLUTING *gave the evenner Crop
and the cleaner Quondal.*

I

But

But whether the advantage gained was fuperior to the labour and attention beftowed at a bufy time of the year, is a moot point.

By Incident in M 6;—*Peafe fhould be buried deep on a dry foil.*

For the Peabeans which were buried two or three inches deep in the ruts of the cart which diftributed the feed, were much more vigorous than thofe which were left nearer the furface.

In this refpect, therefore, Fluting or Drilling has the preference to fowing over the rough plit; except the Seed-feams be left very deep indeed.

Vegetizing Process.

The whole neglected to be difweeded, until the Peabeans had got too high (owing to the want of Self-attendance).

Vegetable Process.

Began to cut the 7th of Auguft, when the halm and upper pods were quite green; in order to improve the fodder, and to prevent the weeds from fhedding their feed. The Docks were drawn out of the Wads immediately after cutting.

Part were mown with *Sithes*; part cut with *Hooks.*

The *Sithes* made the quickeft and the cleaneft work. Therefore,

Perhaps; Mowing *is the moft elegible way of cutting Pea-beans.*

Crop.

Near two Field-Jags an Acre. See the Succession.

Quondal.

The parts which were *difcumbered* before Plowing, *and fluted* after it, are the cleaneft.

The

PEABEANS, 1777.

The parts *difcumbered*, but *not fluted*, the next.

The *fpringy parts*,----as foul as couch and feed-weeds can make them.

Wherever the *Peafe* predominated, the Quondal is *clean*;

Where the *Beans* had the afcendency, *foul* :

The land of *Beans alone*, the *fouleft*.   Therefore,

Perhaps ; *Peafe*, by fmothering the weeds, *cleanfe the foil*; *Beans*, by leaving the furface open, *foul it*.   Therefore,

Perhaps; PEASE ALONE *are a more eligible* Fallow-Crop *than* PEABEANS.

For although there was not this year any *obvious* difference between the land of entire Peafe and the adjoining Peabeans, it does not fet afide this inference ; for the Beans were fo few and fo weak, the Peafe bore them to the ground, and prevented their evil tendency.

IN FUTURE,

I will not venture Beans nor Peabeans on a gravelly foil (except by way of Experiment).

I will endeavour to bury the feed of Peabeans or Peafe, from two to four inches deep.

I will not fow Peabeans, as a Fallow-Crop, on land which will produce a Crop of Peafe alone.

MIXGRASS.

MIXGRASS.

1777.

	containing		produced *		laid at		or trusses an Acre
T.		7⅜ Acres.		14 Jags.		12½ Loads.		61
B, C, D.		19½ ——		21 ——		18½ ——		34
M, N. P.		5¼ ——		5 ——		4¾ ——		30
A.		11½ ——		8 ——		6 ——		19
		44⅛ Acres.		48 Field Jags.		41¼ Sale Loads.		

SOIL.

T ; part sandy Loam,—part strong Loam (tenacious, and some-what clayey).

B, C, D ; Clayey Loam.

M, N, P ; Borders (of various soils), and boggy patches.

A ; Clayey Loam.

Nothing therefore can be drawn, decisively, from the soil.

MANURE.

T ; the whole *very well manured* for last year's Crop.

Part of B, C, D, was *manured* for 1776 ; part was not.

M, N, P ; wholly undressed.

Part of A, was *dunged* for the present Crop ; part was not.

* Began mowing 10th July, and finished 18th August. Began carrying 15th July, and finished 23d August. The reasons why some of the Mixgrass Leys were this year cut so very late were, that part of them (A) was of less value than the Meadow and Tare-Barley, which, on account of the wet weather, had stood until they were full grown ; and that part of them (B, C, D) being principally Cow-grass, on a cold, backward soil, they received less damage by standing than the crops which took precedency of them, would have done : and as a proof that the latter (B, C, D) were not cut *out of time*, the Crop was large for the land it grew upon, the Quondals were left perfectly green, and the Hay, being got without wet, sold at an uncommonly high price. The former, however, (A) was considerably injured by standing too long, though cut ten days before B, C, D.

Therefore ;—*Actual Observation alone can determine the proper time of cutting grass as well as grain.*

By

MIXGRASS, 1777.

By Experiment, No. 22 ;---*Laying dung* on the furface *of a clayey Loam*, in December, *was of no benefit to the Mixgrafs.*

By Experiment, No. 23 ;—*Spreading Compoft*, in December, *over Mixgrafs on a clayey Loam, was of no fervice to the fucceeding Crop*. But,

By Incident in T 3 ;---Perhaps, *Spreading Compoft over young Mixgrafs*, in January, *was of fervice, not only on a fandy, but on a weak, clayey Loam.*

This Incident, however, is not decifive, no comparative Experiment having been made in this field. Befides, although the Crop was large, and the Herbage uncommonly fine, thefe circumftances might be owing to the high tilth of the Soil, and to the pafturing of the Crop the firft year. Be this as it may,—part of a field of *clayey loam* (not fo ftrong as A, C) which was fown, in high tilth, with Mixgrafs-feeds, in 1775, which was manured in the *January*, and paftured in the Summer, of 1776, and which was mowed in 1777, gave an exceedingly fine Crop of valuable Herbage. Therefore, manuring the *furface* of a *clayey Loam* in *Winter* may require more Experiments.

S E E D.

T ; a mixture of *white Clover*, *Ribgrafs*, and *Trefoil*.

B, C, D ; a mixture of *White Clover*, *Trefoil*, *Ribgrafs* and *Cow-grafs*.

M, N, P ; various.

A ; Stable-loft *Hay-Seed*, collected in London, Croydon, &c.

The crop from the *Hay-Seeds* very bad : That from the mixture with *Cow-grafs*, much better than could be expected from the tilth, heart, and fpecies of the foil it grew on. And,

Perhaps ;

Perhaps ;—*For a temporary Ley* (of three to seven years old) *Cow-grass is a most valuable plant.*

The Crop from the *fine Mixture* very good ; and the Herbage well assorted, and full of *Blade*-Grasses. And,

Perhaps ;—*For a* PERPETUAL LEY,—*a* MEADOW,—White Clover, Ribgrass *and* Trefoil *are most valuable* Leaf Grasses.

It must be remembered, however, that T 3. abounds with Docks; which probably were sown with the Grass-seeds.

Therefore ;---*It is very essential to attend to the Pureness of the Seeds of Ley-Grasses.*

WEATHER.

See CLOVER and TARE-BARLEY.

SUCCESSION.

T ; of the second year's growth, pastured the first.

B, C, D ; of the third year's growth, mown every year *.

A ; of the first year's growth.

T, was sown with Barley, after Beans, or after an indifferent Summer-fallow.

B, C, D, with Oats, on one plowing ; principally after Wheat.

Part of A, was sown with Oats, after two Summer-fallows ; part with Oats, after Wheat; part with Oats, after Oats.

After Fallow, thin, (chiefly owing to the badness of the *Hay-seed*, and the dryness of the season of sowing) but beautifully clean. After Wheat, a somewhat better Crop, (the land was in better heart) but very foul. After Oats,—not worth mowing; though part of it was dunged.

* These fields lie inconveniently for Pasturing, and Hay has lately bore an uncommonly high price.

SOIL-

SOIL-PROCESS.

From this Analyfis of the *Succeffion*, it is evident that no pofitive Inference can be drawn this year with refpect to the *Soil-Procefs* proper for Mixgrafs: For A 2. which had been fallowed two years fucceffively, was not fo large a Crop as A 1. which was fown on one plowing. It muft be obferved, however, that the *grafrinefs* of A 1. added to, or in a great meafure conftituted the Crop; and as very few of the feeds which were fown vegetated, it was impoffible that a foil fo perfectly clean as A 2. is, could afford much herbage.

T 3. is a good evidence that Mixgrafs affects fine Tillage. And,

By Experiment, No. 3;—*Mixgrafs was good in proportion to the finenefs of the Tilth.*

SEED-PROCESS.

A decifive Experiment was made in A 2. (See MINUTE of 21 April, 1776.) Each alternate ridge was hand-raked after fowing. The Ridges which were raked are obvioufly better furnifhed with Plants, (the feafon very dry, and the Seeds, by raking, became more deeply buried and better covered), and their furfaces are fmoother than thofe only rolled without raking. Therefore,

Perhaps;—*It is good Management to make Raking in the Seeds of Ley-Graffes a general Practice.*

For although the Benefit to A 1. and 3. was not fo obvious as to A 2. this was owing principally to the Foulnefs of their Surfaces, which rendered the Benefit lefs perceptible. See the MINUTE above referred to.

VEGE-

MIXGRASS, 1777.

VEGETIZING PROCESS.

Part difweeded of Docks and Thiftles, during the moift weather in June. Part was too high to go-over it without injuring the Crop.

VEGETABLE PROCESS.

See a Note, page 39.

T, being cut while the ground was cold and damp, the Herbage was wadded immediately after the Sithe, and frequently turned on to frefh ground.

A, B, &c. having ftood until the Sap was in fome degree diffipated, and being cut while the ground was hot and dry, the Hay, after having lain a day or two to wither, was broken into Beds, turned, &c. and carried without Cocking.

The whole was got into Stack in very high condition.

CROP.

Various : and the circumftances attending the produ&ion of it fo intricate, that no pofitive, nor even probable, *general Inference* can be drawn.

QUONDALS.

Promifing or unpromifing in proportion to the ftates of the different Soils at the time of Leying ; except part of T 3, which looks as brown as a Fallow. The largenefs of the Crop and the wetnefs of the Seafon have not only tainted the lower part of the Herbage, but, I am afraid, injured the Roots.

IN FUTURE,

I will not fow Ley-Graffes over one *Plowing.*

I will not fow common* *Stable-Loft Hay-Seed for a Ley.*

I 2 *I will*

---

* Perhaps, if a well-herbaged Meadow were carefully difweeded, and fuffered to ftand until its Seeds were *properly* matured; the Hay well-got, kept feparate, not too

much

MIXGRASS, 1777.

I will ever hand-rake-in the Seeds of Ley-grasses.

I will not spread Manure on the surface of a retentive Ley when the Soil is wet.

I will not mow *a Ley the* first *Year.*

I will ever disweed a Ley before or after cutting.

I will not suffer a young *Ley to stand until it be over-grown, let the* Weather *be what it may.*

much heated in the Mow, and carefully thrashed ; the *Hay Seeds* thus produced might be very valuable as Grafs-Seeds for a perennial Ley ; especially for a Soil similar to that of the Meadow by which they were produced.

W H E A T.

W H E A T.
1777.

L. ⎤ containing ⎧ 24 Acres ⎫ produced ⎧ 71 Jags. ⎫ laid at ⎧ 75 Quarters. ⎫ or bushels an Acre. ⎧ 25.
S. ⎬ ⎨ 1 —— ⎬ ⎨ 2 —— ⎬ ⎨ 3 —— ⎬ ⎨ 24.
P. ⎦ ⎩ 18½ —— ⎭ ⎩ 33½ —— ⎭ ⎩ 45 —— ⎭ ⎩ 20.

43¼ Acres.　　106½ Harvest Jags.　123 Quarters.†

S o i l.

L, clayey Loam, with a retentive Subsoil,
S, sandy Loam, with a retentive Subsoil,
P, gravelly Loam, with an absorbent Subsoil.

The stiff land produced the best Crop; but it was best-tilled and best-manured: there was very good Wheat on some of the lighter Soils; especially on the sandy Loam, which was in high Tilth and good heart. And,

Perhaps;—*Wheat affects almost every* Species *of Soil.*

M a n u r e.

L, was *dunged* for this Crop.
S, was *very well dunged* for 1776.
P, was *manured* in part for this,—in part for 1776.

* Began Reaping the 12th of August, and finished the 2d of September: Began Carrying the 26th of August, and finished the 8th of September.

† Low as this Estimate may seem, it proved to be above the Truth; the whole Yield being only 903 Bushels of Head, and 60 Bushels of Tail; amounting together to 120 Quarters and 3 Bushels. Where the Crop was large and much lodged, I laid it at a Quarter each Jag; but I apprehend it did not yield so much: whereas in a *yielding Year*, a Jag of equal size to those alluded to will afford from two Quarters to twenty Bushels of Wheat. Such is the pernicious effect of a cold, wet Summer!

I

By

WHEAT, 1777.

By Experiment, No. 17;—*Dung and Compost were equally unserviceable as Top-Dressings for Wheat.*

It must be observed, however, that the Patch on which this Experiment was tried is a tenacious Loam, lying on a cold, retentive Subsoil.

SEED.

Part was sown with Seed raised on a *kindred* Soil.
Part, raised on an *alien* Soil.

By Experiment, No. 6;—*From* Clay to Clay *gave a ranker and a forwarder Crop of Wheat than from* Gravel to Clay.

By Experiment, No. 18;—*From* Gravel to Gravel *gave a forwarder Crop of Wheat than from* Clay to Gravel.

But this has been an unfavourable year for making Experiments in; and I am by no means *satisfied* with these Results: the Corn being every where lodged and ravelled, it was impossible to form a a decisive judgment. The circumstance of *Forwardness*, however, was as obvious as it is remarkable. Nevertheless, the CHANGE OF SEED is an important subject, which calls for future repeated Experiments.

WEATHER.

See the Articles CLOVER and TARE-BARLEY.

This summer is a positive evidence that Wheat does not affect wet weather. Even where the straw *stood*, the grain is shrivelled and light: where the straw was much *lodged*, the grain is uncommonly thin, and the ears almost empty *.

* See Note (†), page 45.

SUCCESSION.

SUCCESSION.

Part was after *Summer-Fallow* ;
Part after *Fallow-Crops* ;
Part after *Clover-Ley* ;
Part of the Fallow-Crops was *Potatoes* (S) ;
Part *Tare-Barley* (Part of L 1.) ;
Part Mazagan *Beans* (the chief part of L 2. and part of L 1. See Experiment XI.)

The *Summer-Fallow* the beſt CROP, and the cleaneſt QUONDAL.

Potatoes, the next.

Part of the *Beans* (ſown early) follows next.

Part of the *Beans* and the *Tare-Barley* (ſown a month later) next ſucceeds in point of CROP ; but the QUONDALS both of the *Beans* and *Tare-Barley* are very foul in compariſon to the *Summer-Fallow*.

The *Clover-Ley* a tolerable CROP ; but the QUONDAL is intolerably foul.

By Experiment, No. 8 ;—Summer-Fallow *gave a larger Crop and a cleaner Quondal than* horſe-hoed Beans.

By Experiment, No. 11 ;—*A good Crop of* Wheat *may be had after a* Summer-Fallow ;—*or after a* beaten Road ; *or* virgin Earth ; *or* Beans ; *or* Tare-Barley.

See more relative to the *Succeſſion* proper for Wheat, in the *General Obſervations*, at the cloſe of this Article.

SOIL-PROCESS.

Part *Summer-fallowed.*
Part *Dog-Days fallowed.*
Part *once plowed.*

By

WHEAT, 1777.

By Experiment, No. 8 ;—*A* Summer-Fallow *of six plowings gave a larger Crop and a cleaner Quondal than a* Fallow-Crop, *with two hoeings, a weeding, and five plowings.*

By Experiment, No. 12 ;—*A* beaten Road *acrofs a Fallow, with five plowings, gave a* ftronger *Crop than a full* Summer-Fallow *of feven plowings :* Therefore,

Perhaps ;—*Rolling Fallows very hard is good management.*

By the fame Experiment ;—*One additional* Crofs-plowing *increafed the Crop confiderably.*

By the fame;—*Four plowings after* Beans twice hoed, *and fix plowings after* Tare-Barley, *gave* weedy Quondals,—*compared with* Summer-Fallow.

By Incident in L ;—*Through the whole divifion, the* Crops *and the* Quondals *were nearly in proportion to the* Tillage *expended.*

The retentive Soils were univerfally acclivated in five-bout beds, and thoroughly crofs-furrowed. The abforbent Soil was part *acclivated,* part plowed *flat.*

By Experiment, No. 15 ;—*The advantage of* acclivating *even a* Gravel *for Wheat was glaringly obvious.*

This Experiment was founded on the following THEORY.
Gravel and other *porous Soils* are faid to " *fwallow Mendment* ;" and perhaps it is a fact, that vegetable food may, by heavy rains, be wafhed down below the vegetative Stratum*.

* The vegetative Stratum, however, is undeterminate ; for different Vegetables feed at different depths.

A *porous*

WHEAT, 1777.

A *porous* Soil *plowed flat* becomes an actual *Filter* (Strainer—acts like a *filtering-ftone*), and the fuperfluous rain has no other way to efcape than by paffing *through* this filter-like Stratum, in which Stratum the vegetable food is depofited: and it is at leaft probable, that the water, in its paffage, becomes impregnated with, and carries away, (perhaps entirely out of the fphere of vegetation) fome of thofe particles (be they what they may) which otherwife would have affifted the Crop.

If, therefore, the Soil can be *depofited* in fuch a manner as *to prevent its acting as a filter*, this evil tendency of a porous Soil may be avoided. And if, by rendering the furface *acclivous*, the fuperfluous rain may be caufed to *run off*, inftead of *paffing through*, the plant-feeding mould, the vegetative Stratum no longer acts as a filter.

Befides, perhaps a *Gravel* which lies flat *binds* more clofely than that which is laid up in narrow ridges; which, having free liberty of expanfion, the Soil, perhaps, is fuffered to remain in a kinder, mellower, more genial ftate.

Whether or not this reafoning is juft, the refult was clearly in its favour; for not only the Crop on the *round ridges* was obvioufly better, but the Soil of thofe broke-up obvioufly more mellow * than that of the *wide flat beds*.

This theory may be extended: If a porous Soil, *lying flat* on an abforbent Subfoil, acts as a filter, it may, with fome degree of reafon, be faid, that fuch a Soil fo fituated has an advantage over one which is tenacious and retentive, and confequently over a fimilar Soil fharply acclivated; for, by acting as a *Filter*, thofe genial particles with which Rain-Water *is faid* to abound, are retained in the vegetative Stratum; while thofe which fall on a lefs porous Soil, or

---

* *Mellow* and *kind* may not be epithets ftrictly philofophical; but in this inftance they convey the Writer's meaning, as an Agriculturift, better than any other which occurred to him.

on

on any Soil *lying in high round Ridges,* are not only hurried away themfelves, but are accompanied by others which happen to lie on the furface of the Soil.

May we not reconcile thefe two modes of reafoning, feemingly oppofite, by drawing from them, jointly, the two following probable Inferences?

Perhaps;—*A porous Soil, fituated on an abforbent Subfoil, and which has been* recently dunged, *or is otherwife* replete with vegetable food, *fhould be depofited in* round Ridges ; *left, by fuffering the fuperfluous Rain-Water to drain* through *the Soil, its riches may be removed from the fphere of vegetation.* But,

Perhaps ;—*A porous Soil, fituated on an abforbent Subfoil, and which has been* exhaufted *by a fucceffion of Crops, fhould be depofited in* flat Beds, *in order that the plant-feeding Mould may admit and retain the vegetable aliment which falls in Rain-Water.*

The patch on which the above-mentioned Experiment was more particularly made, is a fharp gravelly Loam, which had been fummer-fallowed and *dunged,* and with which the Dung had been thoroughly compofted.

By Incident in L.—*Three and half large Jags of Wheat can grow on land laid-up in half-rod Ridges.* And,

Perhaps ;—*Half-rod Ridges are the beft of all poffible Beds for a Clay, or for a Gravel.*

If the *Gravel* be *in heart,* the lands may be laid-up *round;* if *out of heart,* they may be plowed *flat.*

MANURE-

WHEAT, 1777.

MANURE-PROCESS

Part *plowed in*;
Part *laid on* the Surface.

By Incidents in L, P.—*Dunging the rough Plit of* one deep plowing, *and afterwards harrowing and* plowing in *the Dung, is very good management* †.

By Experiment, No. 14;—*Manure*, laid on *a strong Loam before fluting for Wheat, had not the expected effect.*

By Experiment, No. 17;—*Dung and compost were equally useless as* Top-dressings *before fluting for Wheat.* And,

Perhaps;—*Except the Soil be very absorbent*, Top-dressing, *in any manner*, for Wheat, *is bad management.*

Because *Wheat* feeds *deep*; and a *top-dressing*, perhaps, serves only to make it winter-proud. (See a Note to Experiment, No. 14.)

---

* MELIORATION is the most indefinite department of Agriculture, and the most difficult to reduce to Science. MELIORATIONS may be divided into MANURES and STIMULATIVES, each of which are various; and the varieties of Processes appertaining to each are numerous. MANURES, it is true, aptly, and indeed necessarily, follow the SOIL; but the *Processes* and *Stimulatives* may be intimately connected with the *Soil-Process,* the *Seed-Process,* the *Vegetizing Process,* or even (as in the case of *herbaceous Melioration*), with the *Vegetable Process.* The Writer has therefore hitherto treated of MELIORATION at large, under the head MANURE; and his reason for deviating in this Article is, because the *Manure-Process* happened to fall appositely, tho' not distinctly, between the *Soil-Process* and *Seed-Process;* and he embraced it as a favourable opportunity of explaining himself, in some measure, on this interesting subject; and of elucidating, in some degree, this obscure department of the Science of Agriculture. And he embraced this opportunity the more willingly, as he wished to treat the article WHEAT, 1777, as fully and comprehensively as possible.

† See a MINUTE of the 21st July, 1776.

SEED.

WHEAT, 1777.

SEED-PROCESS.

Time of Sowing.

Began sowing the 29th of September, and finished the 7th of November.

One side of L 1. was sown, from three weeks to a month, before the other side.

The *early-sown* was much the largest, rankest Crop; but it was almost wholly *lodged*, and the Grain very light in the ear: Whereas the *late-sown*, in general, *stood*; the ears were large and well filled; and, although the Crop in the *Field* was not more than two-thirds so bulky as that of the *early-sown*, I am of opinion that in the *Barn* the *late-sown* will prove the best Crop.

This equality, however, is merely a casualty of the weather. Had the Summer proved moderately *dry*, the *early-sown* would have been considerably the best Crop; its plants in the Spring were far more numerous and healthy than those of the *late-sown*: And indeed, generally, *the Time of Sowing* is one of those mysteries of Agriculture, which being in some degree dependant on *Chance*, cannot be nicely regulated by human foresight. There may, nevertheless, be one GENERAL RULE FOR THE TIME OF SOWING; which, *taken in a general Sense*, may, *perhaps*, be applicable to every Crop, and to every Country.

Perhaps;—*Sow poor Land early; rich Land late.*

For if the Summer prove *wet*, a field which is out of heart runs no risque of being injured by Rankness, and the field which is full of Manure will be prevented from lodging.

If the Summer prove *dry*, a field which is poor, and which does not get its surface shaded before the drought set in, is in danger of

being

being ftinted, or wholly burnt up ; while a field (of Wheat at leaft) which is in heart, will force its way, in defiance of the drynefs of the weather.

Early and *late*, however, when. applied to *the Time of Sowing*, may each of them have a diftinct meaning in different countries. And indeed not. only every country, but every county, nay, every diftrict, may have, with ftrict propriety, its *peculiar Time of Sowing*. However, as a *general Regulation of the above Maxim*, we may venture to fay,

Perhaps ;—Begin *with the Soil which is* poor, *and* finish *with that which is* in heart.

Preparation of the Seed.

Part of it was *prepared* by fteeping it in ftrong Lime-water, falted fufficiently to bear an egg ; and afterwards limed.

Part was fown *without Preparation.*

By Experiment, No. 5 ;—*Pickling the Seed* feemed. *to be difadvantageous to the Crop.*

By Experiments, No. 7 and 19 ;—*There was not the leaft advantage arofe this year from* Brining *Wheat.*

This is the fecond year I have made Experiments on *pickling Wheat*, without one inftance in its favour. It has always happened, however, that the *entire Pieces* on which the Experiments have been made, were *wholly* free from *Smut* (the difeafe intended to be guarded againft) ; and confequently no *comparifon* could be made. More Experiments are therefore neceffary to a final decifion.

Mode

WHEAT, 1777.

Mode of Sowing.

Part was sown *under-plit*.
Part, over the *fresh Plit, rough*.
Part, over the *fresh Plit, fluted*.
Part, over the *stale Plit, fluted*.

By Experiment, No. 9;—2⅛ *Bushels of Wheat sown over-plit, gave a better Crop, and a cleaner Quondal, than the same quantity of Seed sown* under-plit.

It muſt be obſerved, however, that this Experiment was made in L 1. a *clayey Loam*: on a *light ſandy Loam*, the reſult might have been different; perhaps the reverſe. With reſpect to a ſtiff, cold ſoil, however, this is a very deciſive Experiment: the part ſown under-plit had not half the number of plants as had the part ſown over-plit: and, generally,

Perhaps;—*On cold wet Land, two Bushels of Wheat sown* over the freſh *Plit, is an equivalent to three Bushels plowed in*.

By Experiment, No. 10;—*It was immaterial whether the Soil was harrowed, or left rough, after sowing under-plit*.

(This Experiment, however, is not ſufficiently deciſive.)
A comparative Experiment was made between the *fresh Plit rough*, and the *fresh Plit fluted*; but the whole was ſo lodged and ſo ravelled, the reſult was dubious.

The *ſtale Plit fluted* was plowed when the Surface was covered two or three inches thick with the *third* Crop of Clover; the *freſh Plit fluted* was plowed when the Surface was quite bare, the After-graſs having been paſtured off very cloſe.

By Experiment, No. 13; —*It is better to flute the* ſtale Plit plowed clovery, *than the* freſh Plit paſtured.

4   The

The Experiment on *fluting the ftale Plit plowed clovery* was repeated in two or three different places, and the refults were uniformly in its favour.

Quantity of Seed.

Various: On a par, about 2⅛ Bufhels an Acre; and the Crop in general too rank: but the Seed was principally fown over the *frefh Plit*, or *frefh Flutes*; and,

Perhaps;---*Two Bufhels of Wheat fown over* a frefh Surface, *is equal to two and half Bufhels fown on* a ftale Plit.

A tenacious Soil is here more particularly fpoken of; the furface of which, when newly plowed, abounds with cells and fiffures, which readily receive the Seed; but which are fhut or filled up by the firft fhower of Rain, or even by the Dews and the mouldering of the Soil; and when once thefe *Seed-cells* are clofed, and the Surface has acquired a *Varnifh,*—*a glazen Cruft,* it becomes difficult to cover the Seed effectually.

By Experiment, No. 16;—THIN SOWING *of Wheat on a* GRAVEL *is* fortunate, *when the Summer proves* wet.

The 1½ Bufhel (fee the Experiment) was quite a rank Crop; the 2½ Bufhels, a middling Crop; but the four Bufhels, not more than eighteen inches high; many of the ears not an inch long, and the ftraws not thicker than the ftems of Rye-grafs! But, perhaps, had the Spring and Summer proved *dry,* the firft would have been burnt up; while the laft, by fhading the Surface, and thereby keeping the Soil cool, might have been a good Crop. It muft alfo be obferved, that this Experiment was made over *Flutes,* and probably almoft *every Grain vegetated.*

From

WHEAT, 1777.

From this year's experience, and from repeated obfervations in dry Years, I am convinced that *the Quantity of Seed* for a *burning Gravel*, cannot be *nearly* afcertained without a fore-knowledge of the Weather of the enfuing Summer.

Therefore ;---*Burning Gravels are hazardous Soils :*

Becaufe the Crop depends effentially on *the Quantity of Seed* ; and the *proper* quantity of Seed depends wholly on that *Weather* which cannot be forefeen : therefore,

Perhaps ;---*Sow on a Gravel from* 2½ *to three Bufhels of Wheat an Acre ; and, if the Winter prove* dry, *thin the plants with a Hoe in the Spring; but, if the Winter prove* wet, *let the whole ftand in expectation of a* dry Summer *.

Covering.

The whole, whether Fallow or Ley, (except a part fown under-plit) was harrowed as fine as a Garden. No labour was fpared until the beds were rendered (by the concave Hinge Harrows) *perfectly convex,* their *Surfaces fine,* and the *Seed covered.*

Adjufting.

Part of the Inter-furrows were opened with the double Plow ; part left cloddy ; and a comparative Experiment was regiftered ; but the whole was fo rank and fo lodged, no accurate inference could be drawn. The wet Soils were carefully crofs-furrowed, fufficiently *deep* to drain effectually the Inter-furrows, and fufficiently *wide* to walk in.

* But fee a Note to Experiment 43.

V E G E-

WHEAT, 1777.

VEGETIZING PROCESS.

Part of the Inter-furrows were intended to have been Ox-hoed; but not embracing the tranfient opportunity, the Plants got too high for the operation. *Perhaps*, had the Inter-furrows, throughout, been properly kept open, the Crop would have been confiderably benefited. See MINUTES of 16 May, 1775, and 17 Auguft, 1776.

The whole was carefully difweeded.

The part Summer-fallowed, being very rank in the Spring, I experimentally mowed one Land, in the wane of April, to try the effect of checking the Crop.

By Experiment, No. 43 ;---*Mowing down rank Wheat, in April, the* Summer *proving* wet, *was beneficial to the Crop.* Therefore, in future,

Perhaps ;---*If the* Winter *prove* dry*, *pafture or mow down rank Wheat, in March-April.*

But it muft be obferved that,

By Experiment, No. 43 ;—*Mowing Wheat in April fouled the Soil.* Therefore, in future,

Perhaps;---*If* Winter *prove* wet, *let Wheat which is rank, take the chance of the Summer.*

For, beyond a doubt, *checking the Crop fouls the Soil, leffens the quantity of ftraw, retards the ripening*, and, in cafe of a *dry Summer*, muft injure the Crop very confiderably : Therefore, in future,

Perhaps ;---*Whether the Winter prove* wet *or* dry, *do not check Wheat which is not very rank.*

* See a Note to Experiment 43.

L

However,

WHEAT, 1777.

However, if Wheat in March-April fhew an extraordinary inclination to ranknefs, it is obvioufly good management to check it;
not only as a preventive againft *lodging*, but by way of gaining
a valuable acquifition of SPRING FEED, whether VERDAGE or
PASTURAGE.

VEGETABLE PROCESS.

Began to reap the 12th of Auguft.

The Crop in general was fo very large, and fo exceedingly lodged,
I almoft defpaired of getting it cut. Two Kentifh-men (in general
good reapers) began to reap the worft part of it at four fhillings
a-day.—They did not cut more than twenty rods! but they bound at
the rate of 100 fhocks (of ten fheaves) an Acre! A patch fome-
thing lighter was then let at fixteen fhillings an Acre; but the men
could not earn more than half-a-crown a day (which is confidered
as a poor pittance in harveft); I therefore raifed the price to eighteen
fhillings: But even this would not do; and the major part of the
24 Acres in L. was cut at twenty Shillings an Acre! One fett gave
out becaufe I would not give them a Guinea an Acre!

Had not this Wheat come nearly a week earlier than the Wheats
in general of the neighbourhood, it would not have been reaped even
at thefe prices; but hands were plentiful, and each fett was glad to
take a fmall piece by way of making a beginning; and,

Perhaps;---*On a par of years, early-ripe Wheat may be
reaped the cheapeft.*

The following precautions were given to each fett of reapers:

" Be careful not to make the fheaves too large:

" Tye them fecurely, but not too tight:

" Let the *wreaths* (or *twifts*) of the bands be turned *upward*,
toward the ears; that the *tails* (or *butts*) of the bands may hang
down, to convey the rain-water on to the ground.

3

" In

WHEAT, 1777.

" In fetting up the fheaves, be careful to place the *ears* of the bands *inward*, and do not croud the fheaves too clofe to each other in the fhocks."

THE CROP.

On the whole, very large : much too large ; being in general very much lodged. Off 43½ Acres we carried 106½ middling Field-Jags ! Sixteen Acres of the rankeft part of L. produced 58½ Loads : upwards of 3½ Jags an Acre !

QUONDALS.

After the parts *summer-fallowed,* the Quondals are *beautifully clean.*
After the parts *fallow-cropped, middling.*
After the *Clover leys, very foul.*

It muft be remembered, however, that thefe Clover-leys were after Barley, after Wheat, &c. with only a Winter and Spring Fallow ; which, notwithftanding the labour beftowed upon it, was not adequate to the tafk of cleanfing the foil fufficiently for Spring-corn and Clover.

GENERAL OBSERVATIONS.

What was the management of the fixteen Acres which this year produced at the rate of 3½ Field-Jags an Acre ? The Soil, a clayey Loam, was part of it a Summer-fallow ;--part a Bean-Quondal, Dog-days fallowed : the whole dunged with about ten fifty-foot loads (about 500 cubical feet) of prime horfe-dung an Acre, fpread over the rough Plit of *one deep plowing**; harrowed ;--rolled ;--gathered into half-rod ridges *very fhallow* ;--harrowed ;--rolled ;- the ridges reverfed *moderately deep* ;--fown over the frefh Plit, in very high Tilth, in September-October;--harrowed extremely fine;--the Inter-furrows opened, and the Crofs-furrows made wide and deep.---Although it is very flat, wet land, not a fpoonful of water ftood on it during the Winter.

* See MINUTE of 21 July, 1776.

The

WHEAT, 1777.

The avocation of Agriculture would indeed be difheartening, if a good crop of Wheat could not be obtained from fuch management, and fuch weather as attended thefe fixteen Acres. And, were Autumns in general as favourable as the laft was, I fhould almoft give up the thought of *Wheat on a Clover-Ley* : but fuch another Wheat Seed-time may never happen.

Had the divifion L. been caught in a *wet Autumn*, one-half of it at leaft could not have been cropt with Wheat, and the Seed of the other half muft have been put in very badly. Fine as the weather happened, the labour, attendance, and attention beftowed on it was without end ; and the anxiety for the weather equal to the difagreeable watchfulnefs of hay-time and harveft.

What would have been the cafe this year, had I had nothing but *Fallows*—even *Summer-Fallows*—to depend on for Wheat ? The weather, from the middle of May to the middle of July, was incefſantly rainy ; and the ample crops of this harveft have fully employed the teams ever fince. Part of K 4, a Summer-fallow, is now (10 September) as green as a Ley ! I have not an Acre of Fallow fowable with Wheat without two or three more plowings. Had I nothing but *Fallows*, I could not, be the weather ever fo fine, put in fifty Acres of Wheat, tolerably, before Chriftmas. Very fortunately, however, I have *Clover-Leys*, moft of them dunged, ready to be landed-up for Wheat, as foon as rain comes to moiften the furface : And my Fallows being for Spring-corn, they will receive the Winter and Spring ftirrings. 1 have no hope of getting fuch noble Crops and beautiful Quondals from *Clover-Leys* as from *dunged Summer-Fallows* ; but I hope that my Spring Crops and Autumnal Comfort will over-balance even the valuable advantages of a good Crop of Wheat and a Clean Quondal.

The Crop next in goodnefs was from a clayey Loam, part of it a Bean, part a Tare barley-Quondal, each of which received a very

I   good

good Dog-days Fallow, and were managed almoft exactly by the procefs abovementioned; excepting that this part of the divifion L., was fown in October-November—That in September-October.

The next which followed in point of goodnefs, was raifed from a Clover-Ley on a gravelly Loam, dunged for Clover; landed up by two Oxen and a whip-rein Plow, foon after the fecond Crop of Clover was off;—lay three weeks in rough Plit;—it was then harrowed,—. fluted,—(fown,—) and harrowed, with one horfe only: the whole expence of labour not five fhillings an Acre.

Laft Autumn this ftruck me as a moft eligible procefs: this Harveft has convinced me that Theory was once right. During Winter and Spring, the Crop was beautiful; and, had it not been lodged by the heavy rains of laft Summer, it would have been a very good Crop at Harveft; and, notwithftanding the wetnefs of the Summer, and the pronenefs of the foil to *Seed-weeds*, this Crop was almoft wholly free from them; for they had vegetated abundantly while the Soil lay in rough plit, and the harrows and the flute totally eradicated them.

The Quondal, it is true, is foul with *Root-weeds*; but this muft not be charged to the difadvantage of the fucceffion of *Clover-Wheat*, nor to the procefs of *fluting the ftale plit of a Clover-Ley*, but to the flovenly fucceffion of *Wheat, Barley, Clover*. A *clean* Clover-Ley, *properly plowed*, cannot poffibly afford a foul Quondal.

IN FUTURE,

I will never depend wholly *on* Fallows *for Wheat*.

A SUFFICIENCY OF *clean* CLOVER-LEYS FOR WHEAT SHALL BE THE PRINCIPAL OBJECT OF MY FUTURE MANAGEMENT*.

* It muft be obferved, that thefe Refolutions are formed in the neighbourhood of London, where Clover-hay generally bears a price equal to that of the beft Meadowhay.

If

WHEAT, 1777.

If the Soil be much out of heart, I will dung one deep plowing *for Spring-Corn and Clover.*

If the Soil be in tolerable heart, I will top-dreſs *for the Spring-Corn and Clover, and* dung *for the Wheat.*

If Auguſt *be moiſt, I will endeavour to* flute the ſtale Plit.

If the Surface remain droughty *until* September-October, *I will ſow on the* freſh Plit.

If the Soil be poor, *and Manure ſcarce, I will endeavour to* bury the ſecond Crop *of Clover.*

I will endeavour *to* BEGIN *ſowing on the* pooreſt, *and* FINISH *with the* richeſt, *Soil.*

I will not brine *the ſeed of Wheat ; except by way of Experiment* *.

If the Crop be inclinable to rankneſs, I will hoe *the* INTERVALS, *and* paſture, *or* top, *or* verdage *the* BEDS.

I will begin to cut while the knots are green ; and endeavour to let the Sheaves have a ſhower in the Field.

* If Seed-Wheat be foul with light Weed-ſeeds, it may be convènient to immerge it in *Water,* in order to gain an opportunity of *ſkimming off the Lights.*

WINTER-

WINTER-TARES.

1777.

The Acre and half in S 1. and half an Acre in O 1. were mown for Verdage, for the cart-horfes, &c.

$2\frac{1}{2}$ Acres in O 3. produced 8 Field Jags, laid at 5 Quarters.

SOIL.

Part, light gravelly Loam.
Part, ftrong gravelly Loam.

MANURE.

Had a fprinkling of Compoft for 1776.

SEED.

Raifed in G 1. a Loam.

WEATHER.

See CLOVER and TARE-BARLEY.

SUCCESSION.

The light part after *Clover* ;—
The ftiff part after *Barley*.

SOIL-PROCESS.

Once plowed.

SEED-PROCESS.

Sown broadcaft, the 22d of October. The quantity of Seed $2\frac{1}{3}$ Bufhels an Acre.

VEGETIZING PROCESS.

Difweeded of Docks.

VEGE-

WINTER-TARES, 1777.

VEGETABLE PROCESS.

Mown the 8th to the 12th, and carried the 12th and 16th of September.

CROP.

Very bulky; especially on the *lighter* part, after *Clover.*

QUONDAL.

Foul, the part after Clover especially; this part having had only two plowings since the year 1774.

GENERAL OBSERVATIONS.

The reasons for drawing out this formal detail for two and half Acres of Tares are these: First, to retrieve, in some measure, the character of *Seed-Tares*; (See MINUTES of 19th September, 1775, and 11th March, 1776.) and next, to shew that a Farmer has, sometimes, a *good Hit:* for a field which *to appearance* required a Summer-Fallow, has produced, by means of the simple process above-mentioned, a Crop worth at least Nine Guineas an Acre; Tares for Seed *happening*, this Autumn to bear the very high price of Half-a-Guinea a-Bushel.

OATS.

O A T S.

1777.

	containing	produced*	laid at	or bush, an Acre
G.	10 Acres.	20½ Jags.	40 Quarters.	32
B 4.	4¼	11½	16	30
A.	5½	15	20	29
B 3.	4½	6	11	20†
	24¼ Acres.	53 Field-Jags.	87 Quarters.	

S O I L.

G ; fandy Loam.

A, B ; clayey Loam.

There were very good Oats on each fort of Soil : And,

Perhaps ;---Oats affect almoft every fpecies of Soil.

M A N U R E.

Part of G, had a *fprinkling of Dung* for the Oats.

The other fields have been exhaufted by three or four Crops. Neverthelefs, the burdens of A and of B 4, *which had been fummer-fallowed*, were uncommonly *large* ; and, where proper Seed was fown, the *Grain* is fine.

Therefore ;---A good Crop of Oats may be procured without any other Melioration *than* Summer-fallowing ‡.

* Began mowing the 26th of Auguft, and finifhed the 6th of September. Began carrying the 2d of September, and finifhed the 13th.

† Had the Seed of B 4. been the fame as that of B 3. the difparity would have been ftill greater.

‡ It muft be obferved, however, that *Tillage*, though a *Melioration*, is only a *Stimulus*, or *Provocative*,---not a *Manure*.

M   S E E D

OATS, 1777.

SEED.

Having frequently been provoked by *the shedding of early Oats*, I was determined to obtain a *less free-shedding* species; and accordingly I procured some *Scotch* Oats, immediately from Alemouth; but they being sown rather too late, and the heavy rains of June and July setting in, they ran very much to Straw, with only a small *pro portion* of Grain.

I also procured some *Scotch* Oats which had been sown two or three times on the chalky hills of Surry. These gave less Straw, but proportionably more Corn; which, considering the wetness of the Summer, is tolerably good. And,

By Experiment, No. 38;---Scotch Oats *which have been sown on* a chalky *Soil, in* England, *are fitter for a* clayey *Loam than a similar species immediately from* Scotland*.

WEATHER.

See CLOVER and TARE-BARLEY.

SUCCESSION.

G, After *Pulse.*
B 4, A, After *Fallow.*
B 3, After *Wheat.*

By Incidents in A, B ;---*Summer-fallowing a foul clayey Loam for Oats and Clover is most* husbandly *management.*

The generous countenance of the Soil, and the evenness and vigour of the young Grasses, render each Field a *Picture!*

* The same, or a like, Improvement of *Scotch Oats* by sowing them in England, is *said* to take place, although the Soil in England be not varied: A *change of Soil*, however, is more likely to encrease than diminish the Improvement.

By

O A T S, 1777.

By Incident in B 3 ;---*Oats and Clover, after Wheat and one plowing, is most* flovenly *management.*

The countenance of the Soil (where its face can be feen) is pallid and fpiritlefs ; and the fcattered Plants of Clover feem to hide themfelves among the *Weeds,* as if confcious of their worthleffnefs.

It is proper to mention, however, that this Field was fown with Oats and Clover, in order to *clafs* it with the adjoining Field : but, inftead of *Clover,* it muft next year be a *Summer-fallow* ; or, if Dung can be procured, a *Fallow-Crop :* The Clover *Seed,* however, is the only *lofs* ; and in lieu thereof, I have *gained* a leffon which I hope I fhall ever remember. For,

By Incidents in B 3, 4 ;---*Summer-fallowing a clayey Loam for Oats and Clover is as fuperlatively* good, *as fowing them on one plowing of a Wheat-Quondal is unpardonably* bad *management.*

S o i l - P r o c e s s.

G 1, was once plowed ; G 2, thrice plowed.

A, had five, and B 4, fix plowings ; with repeated harrowings, rollings, &c.

B 3, plowed once.

The reafon why the divifion G had not more Tillage was, the Oats were not intended to be fucceeded by Clover, but by Fallow, or Fallow Crop ; in order to bring this divifion into a REGULAR SUCCESSION with F, H, I.

S e e d - P r o c e s s.

Sown *broadcaft,* over the *frefh plit,* between the 22d *of March* and the 17th *of April* ; the *quantity of Seed* 5$\frac{1}{3}$ Bufhels an Acre*.

* This perhaps may be thought to be a large quantity of Seed. The Writer is of opinion that 2$\frac{1}{4}$ Bufhels of Wheat, or 2 Bufhels of Barley are, as Seed, equal to 5 Bufhels of Oats. The reafon why this year the quantity exceeds 5 Bufhels is, becaufe fome of them were fown very late.

VEGE-

O A T S, 1777.

V E G E T I Z I N G   P R O C E S S.

The Fields fummer-fallowed were perfectly clean; the other Fields were difweeded.

V E G E T A B L E   P R O C E S S.

Mown between the 26th of Auguft and the 6th of September: carried between the 2d and the 13th of September.

By Experiment, No. 59;—*Oats fhould not be cocked until the day they be carried.*

C R O P.

Various; according to the Species of *Seed,*—the *Succeſſion,*—and the *Soil-Procefs.*

Q U O N D A L S.

Various; according to the *Succeſſion*, and the *Quantity of Tillage:* except G 2, which had three plowings, and which is as foul as G 1, which had only one plowing. This, however, muft not be placed to the difadvantage of Tillage; as G 2, was proportionably fouler than G 1, before they were broke-up.

I N   F U T U R E,

Let the whole courfe of my Stiff-land PLOW-MANAGEMENT *tend towards cleanfing the Soil* effectually *for Oats and Clover.*

I will endeavour to trench *in November and December, and* crofs the entire Ridglits *in May.*

Where the Soil has been much exhaufted, I will endeavour to lay a fprinkling of Dung on the rough plit of one deep plowing *.*

* It muft be underftood, that this *deep* plowing fhould not be done before the Surface has been made *perfectly clean,* by *moderately deep* plowings, and by harrowings, rollings, &c.

3

Where the Soil is in tolerable heart, I will endeavour to sow Soot *with the* Clover-Seed, *and* Dung *for the* Wheat.

Thus a Crop of *Oats* will be secured by the *Tillage* ;—a Plant of *Clover* by the *Soot* ;—and a Crop of *Wheat* by the *Dung :* with a Quondal sufficiently clean, and sufficiently in heart to admit of a *Fallow-Crop* as a preparative to *Oats, Clover, Wheat,* and a *Summer-Fallow.*

THUS, EVERY EIGHT YEARS, THE SOIL WILL RECEIVE, ALTER-NATELY, A FALLOW AND A FALLOW-CROP.

PRODUCE.

PRODUCE.

1777.

45	Acres of CLOVER,	produced 72½ Jags,	laid at 54½ Loads.
17¼	——— MEADOW,	——— 27 —,	——— 25½ ———
0½	——— BARLEY,	——— 1 —,	——— 2 Quarters.
7¾	——— TARE-BARLEY,	——— 14 —,	——— 12½ Loads.
10½	——— PEABEANS,	——— 20 —,	——— 22½ Quarters.
42¼	——— MIXGRASS,	——— 47 —,	——— 40¾ Loads.
43½	——— WHEAT,	——— 106½ —,	——— 123 Quarters.
2½	——— TARES,	——— 8 —,	——— 5 ———
24¼	——— OATS,	——— 53 —,	——— 87 ———
	——— CLOVER, (2d Crop)	——— 24 —,	——— 24 Loads.
194½	Acres.	373 middling Field Jags.	
22¼	——— PASTURAGE,		
2	——— VERDAGE,		
1½	——— POTATOES,		
1½	——— TURNIPS,		
1½	——— CABBAGES,		
16¼	——— FALLOW, &c.		
29	——— GIVEN UP, &c.		
22	——— WASTE, &c.		
291	Acres.		

The MINUTES OF AGRICULTURE reaching down to JULY 1777, and the DIGEST of the MINUTES not being finiſhed before the enſuing Spring, the *general Matter* of the Year 1777 became incorporated with that Work; and is, of courſe, already publiſhed: A GENERAL REVIEW OF 1777 is, therefore, diſpenſed with, or in ſome part poſtponed, — until JANUARY 1779, when the Writer means to take A GENERAL REVIEW FROM JULY 1777, TO THE CLOSE OF THE YEAR 1778.

THE

THE
ARRANGEMENT.
1778.

TARES.

Time of Sowing.		Quantity of Seed.
October 6 to 9.	1½ Acre in S 2.	3 Bushels of O.

RYE.

October 8 and 9.	1¼ Acre in S 4.	4 Bushels.
November 4.	1½ ——— S 5.	4 Bushels.
	3 Acres.	8 Bushels of Seed.

WHEAT.

October 16 to 20.	6¼ Acres in I 2.	14 Bushels of P
20 to 24.	5¼ ——— I 1.	12 ———
November 6 to 12.	11 ——— of O 1.	27½ ——— L.
14 to 16.	4 ——— in N 6.	12 ———
16 to 18.	4 ——— of N 5.	12 ———
19.	1½ ——— in S 3.	4 ——— P.
	32½ Acres.	81½ Bushels of Seed.

PEASE.

February 12.	3 Acres of P 1, 2.	10 Bush. Marl. greys.
March 12.	10 ——— in H.	30 ——— white boiling.
14.	4 ——— P 3.	13 ——— Marl. greys.
	17 Acres.	53 Bushels of Seed

OATS.

ARRANGEMENT, 1778.

OATS.

24 Acres in L.
6¼ ——— K 4.
30¼ Acres.

BARLEY.

18¼ Acres in M.
9½ ——— in F.
1½ ——— in S 1.
29¼ Acres.

CLOVER.

8¾ Acres in B 3, 4.
4¼ ——— in O 1, 2.
13¼ Acres.

MIXGRASS.

17 Acres in A.
5¼ ——— C.
7¼ ——— D 1.
5½ ——— K 2.
3¼ —— of M 4. 6.
1⅞ ——— N.
1¼ ——— P 1.
7¼ ——— T 2, 3.
49¼ Acres.

MEADOW.

3⅞ Acres in D 2, 3.
13¼ ——— K 1, 3.
1 ——— L 1.
0½ ——— P 2.
4⅞ ——— R 1, 3.
5¼ ——— T 1, 4.
28¼ Acres.

PASTURE.

4¼ Acres in R.

FALLOW

6¼ Acres in B 1, 2.
10 ——— G.
10¼ ——— P 1.
27¼ Acres.
32 ——— given up, &c.
21⅞ ——— Hedges, &c.
291 Acres, including Waste, &c.

IF the Writer had not already explained, in the Advertisement prefixed to this Publication, the disagreeableness of his situation in the Spring of 1778, the foregoing Arrangement would need an Apology; it being in many instances directly contradictory to the Principles of Management which he had the preceding year laid down. It seems, therefore, necessary to observe, that *after the Insolvency* there mentioned, the Writer considered himself merely as an *Agent* or *Middleman* between the *Assignees* and the *Landlords*: And it being determined
that

that the Leafes (they being much too high-rented) fhould be given up, it became proper that the *Improvements* which had been beftowed on the land, and which were now going to be *given-up* with the Leafes, were to be applied, as much as poffible, to *prefent* ufe, without looking forward to *future* benefit ; provided the Farms (excepting one) were *left* in *as good* a ftate as that in which they were *found*.

On thefe principles, twenty-feven Acres of *Summer-Fallow* may feem a large portion : thefe Fallows were broken-up in Winter, when the Writer had not the moft diftant idea of what was to happen in the Spring: they were of courfe *foul*, and many of them were *much out of heart*. To have cropped them in the Spring would have been of very little advantage to the Affignees, and would have been exceedingly injurious to the Farms ; they were therefore each of them (except two fmall Fields) *plowed three times, with intermediate harrowings, &c.* To have done more would have been exceeding the limits of *juftice* ; to have done lefs, would have been injuring the Farms *

* Should an Agent of one of the Farms think his Principal hardly ufed by this line of conduct, his conjectures muft proceed from a total want of agricultural knowledge : *one half* of the Arable Part of the Farm alluded to having been perfectly cleanfed, and brought to an exceedingly high degree of Tilth, either by Summer-fallowing, or by Fallow-Crops in 1777 ; and *the other half* being either Summer-Fallow or Fallow-cropped in 1778. Had the Summer-Fallow been ftirred, and the part Fallow-cropped been broken up, *as foon as poffeffion was given*, and had the other half been cropped with any degree of judgment, this Farm in particular would have fallen into that hufbandman-like, or rather that garden-like ftate, which it was intended for, and into which it muft neceffarily have fallen, (the Summer of 1778 being dry) had not the ordinary courfe of management been unfortunately broken. It would, no doubt, have been *very clever* to have found twenty Acres of Fallow perfectly clean, well dunged, and ready to have been fown with Wheat : But who, with one agricultural idea, could have expected it ? Nay, who could have any *right* to expect to find a Farm in fuch a ftate, as that by *two additional ftirrings of about twelve Acres*, there would not have been *one foul Acre* left in the whole Farm ?

N

From

ARRANGEMENT, 1778.

From an idea of the difagreeablenefs of the fubject, the Writer had almoft determined to fupprefs the Management 1778 ; but on reflecting that he fhould thereby lofe an opportunity of corroborating the Plans of Management he had adopted ;---that the Experiments and Obfervations of 1777 would lofe much of their effect ; and efpecially as he fhould lofe an opportunity of throwing frefh light on the *actual Bufinefs of a Farm*, by giving Reafons for deviating from a regular Plan of Management ; he confidered the idea as unworthy of any man whofe ambition leads him to become a contributor to the ftore of fcientific knowledge.

CLOVER.

C L O V E R.
1778.

B 4.		containing	4¼ Acres	produced	7 Jags	laid at	7 Loads.	or Truffes an Acre	59.
O, 1, 2.			4½ ——		5 ——		4 ——		32.
B 3.			4¼ ——		2½ ——		2 ——		16.
			13¾ Acres.		14½ Field-Jags.		13 Sale-Loads.		

S O I L.

B ; tenacious, clayey Loam.

O ; fandy, gravelly Loam.

The Crop from B, where the Soil was Summer-fallowed, (See CLOVER, 1777) was very good ; and even in B 3, the few ftraggling Plants which got foot-hold were remarkably ftrong and vigorous; although neither of thefe Fields have been manured fince 1774*. Therefore,

Perhaps ;----*Clover affeɛts, peculiarly, a tenacious, clayey Loam.*

The 4¼ Acres in O, were made up of fuch patches as had, in Autumn, been deemed too foul for Wheat. (See SUCCESSION.) Nothing therefore can be drawn from them with refpeɛt to the *Soil.*

M A N U R E.

S E E D.

Part *common Red-Clover.* (*Perhaps* the *French* or *Flemifh* fpecies of red Clover.)

Part *Cow-Grafs.* (*Perhaps,* the native *Englifh* Red-Clover.)

* Excepting part of B 3, footed for 1776 :---They were both intended to be dunged for the fucceeding Crop of Wheat.

By

CLOVER, 1778.

By Experiment, No. 44;—Cow-Grafs *and Red-Clover are obviously different species of Plants.*

By the fame;—Cow-Grafs *is preferable for one Crop of Hay and After-Pasture :* Clover, *if two Crops of Hay are wanted.*

By Incidents in C, D, K; Cow-Grafs *will stand four years.*

Therefore;—Cow-Grafs *three or four years, on a Soil it affects, must be peculiarly profitable :*

Especially as it reaches its perfection the first year, and does not, like Lucerne and Saintfoin, require three or four years to bring it to maturity *.

WEATHER.

Autumn was remarkably dry.

Winter, upon the whole, dry, but not much frost : December-January was, for near a month, uncommonly gloomy, raw, uncomfortable weather ; with fome fhowers of fnow, fleet, and rain.

Spring, feafonable, but inclinable to dry ; with fome charming rains in May.

---

* The Cow-Grafs giving no hope of a *fecond Crop,* it was *paftured* (and gave an abundance of Feed); except an experimental ftripe in the middle of a Field of red Clover, (B 4.) which being exceedingly clean and a promifing Crop, I fuffered to ftand. for *Seed.* This gave the Cow-Grafs a fair opportunity of fhewing itfelf. When the Clover Seed was ripe, the Cow-Grafs was but coming out into head, and to get it out of the way before the Seed Clover was cut, it was mowed and made into Hay, about the middle of September; and produced at the rate of about a load. an Acre. But the firft Crop was cut (on account of the Clover) before it was ripe, and the middle o September is too late in the year to make Hay.

Therefore ;—Cow-Grafs *cannot with propriety be cut more than once.*

4

Summer,

Summer : Hay-time remarkably fine ;—between which and Harveſt we had a fine rain.—Harveſt free from rain.

Clover in general this year was not *tall,* but very full of *Leaf* and *Head.* Laſt year it was all *Stem.* Laſt year, May-June was remarkably wet ; this year, inclinable to dry, with ſome ſhowers :

Therefore ;---*Clover affects* moiſture *rather than* wet.

SUCCESSION.

B 4, after *Oats,* after *Summer-Fallow.*
O, after *Clover,* after *Clover,* after *Barley.*
B 3, after *Oats,* after *Wheat.*

The part *Summer-fallowed,* beyond compariſon ſuperior to that ſown on one plowing after *Wheat :* the produce of each Field was not taken down ſeparately ; but, from memory, B 4, produced ſeven ; B 3, two loads ; and the quality of the former at leaſt one-fourth ſuperior. Added to this, the Quondal of B 4, is uncommonly beautiful ; while that of B 3, is foul with Seed-Weeds of every ſpecies *.

Therefore ;--*Fallowing for Clover is moſt eligible management.*

As to O 2, and the patches in O 1, they are as foul, and the Crops as bad as could poſſibly be expected from the third year of Clover originally ſown on an indifferent Winter-Fallow. They were ſuffered to ſtand this year, becauſe they were too foul to be ſown with Wheat ; and, in regular ſucceſſion, the diviſion O would have

---

* This Field was intended for, and would in the ordinary courſe of things have been Summer-Fallow. But happening not to be broke up in Winter, it was thought *proper* to let it take its chance for a Crop. And it has at leaſt afforded a ſtriking leſſon not to expect a Crop of Clover after one plowing, except the Soil be in extraordinary heart.

been

CLOVER, 1778.

been Summer-fallowed, or fallow-cropped, (as its different Degrees of foulness might require) in 1779, for Barley and Clover in 1780 and 1781.

SOIL-PROCESS.

See the SUCCESSION; from which it is evident that Clover requires good Tillage.

SEED-PROCESS.

The Seed hand-raked in; neverthelefs B 3, almoft wholly miffed, and was refown in Summer. But,

By Experiment, No. 55 ;—*Refowing Clover, when the Oats were in Haw, was of no perceptible fervice.*

It muft be obferved, however, that this Field was only once plowed for the Oats and Clover, and was much out of heart.

VEGETIZING PROCESS.

Swept and rolled.

VEGETABLE PROCESS.

Mown the latter end of June, when the *Clover* was *in full head:* But the experimental flips of *Cow-Grafs* being a fortnight or three weeks backwarder, they were barely *in bud.* The whole was wadded,—turned,—had fome fhowers ; but was carried in good order.

CROP.

After *Summer-Fallow*, near two Loads an Acre.

After *one plowing,* not half a Load an Acre.

QUONDALS.

CLOVER, 1778.

QUONDALS.

After *Summer-Fallow* perfectly clean, and as fit to be sown with Wheat as a Summer-Fallow can possibly be.

After *one plowing*, as foul as Weeds can make it, and fit for nothing but a Summer-Fallow.

IN FUTURE,

I will not venture Clover after any thing but a clean Fallow.

If After-pasture *will be wanted, I will sow* Cow-Grass; *if* Hay, Red-Clover.

MIX

MIXGRASS.

1778.

	containing		produced		laid at		or Truffes an Acre
A 4, 5.		5¼ Acres.		13 Jags.		11 Loads.	72.
T 2, 3.		7¾ ——		10 ——		9 ——	44.
M, N, P.		6½ ——		5½ ——		5 ——	28.
K 2.		5¼ ——		5 ——		4 ——	27.
A 1, 2, 3.		11½ ——		10 ——		8 ——	25.
D 1.		7¾ ——		5 ——		5 ——	23.
C 1, 2.		5¼ ——		2 ——		2 ——	14.
		49¾ Acres.		51½ Field-Jags.		44 Sale-Loads.	

SOIL.

Various: but the Crops feemed to be influenced by the *Seed* and the *Soil-Procefs*, rather than the *Soil*.

By Experiment, No. 45;—*A tenacious clayey Loam is better affected by* Cow-Grafs *than it is by the* finer Graffes *.

I mean, *Cow-Grafs* affords a larger burden the *firft Year*; but, perhaps, the *fine Mixture*, by fuffering the *spontaneous Herbage* to exert itfelf, would fooner afford a *Meadow*.

MANURE.

See OBSERVATIONS 1777, Page 39, and the EXPERIMENTS there referred to. No Manure has this year been laid on Mixgrafs-Leys.

SEED.

Mixgrafs is an indefinite term; and it is here meant to reprefent *any Mixture* of Grafs-Seeds which is fown with an intention to propagate a *perennial Ley*,—a *Meadow*.

Perhaps, each fpecies of *Soil* requires a diftinct *Mixture* of Grafs-Seeds: for,

* White Clover, Trefoil, and Ribgrafs.

By

By Incident in T 3 ;—WHITE-CLOVER, RIBGRASS, *and* TREFOIL, *are a moft excellent Mixture for a* Loam, *or* fandy Loam.

By Incidents in A 1, 2, 3;---HAY-LOFT SEEDS *are unfit for a* clayey Loam.

By Experiment, No. 45 ;—WHITE-CLOVER, RIBGRASS, *and* TREFOIL, *were not fo good* the firft Year, *on a* tenacious clayey Loam, *as was the fame Mixture with an addition of* COW-GRASS.

This, however, is no proof that *Cow-Grafs* is eligible as a *perennial Ley-Grafs* : on the contrary,

Perhaps; — *The more luxuriant the* alien Herbage *is the* firft Year, *the longer it will be before the* native Graffes *form a* meadowy Sward.

WEATHER.

Dry, but not unfeafonable : See CLOVER, page 76.

SUCCESSION.

See the OBSERVATIONS of laft year, page 41.

Part of the *thin* Leys were *mowed*; part were experimentally left to be *trampled* down by the pafturing ftock.

By Experiment, No. 56 ;—*There was no obvious difference between the Parts* mowed *and thofe* trampled.

The irefh Leys this year are A 4 and 5 ; which were fown with *Oats* on a *Summer-Fallow.*

The Crop was exceedingly good, and the Quondals beautifully clean.

O   Therefore :

Therefore;—*Summer-fallowing for Spring-Corn and Ley-Graſſes is eligible Management.*

SOIL-PROCESS.

In Autumn, the Fallows were gathered up into half-rod Ridges, and croſs furrowed as carefully as they would have been if ſown with Wheat. In Spring, theſe Ridges were thrown two-into-one; and, after the Oats were ſown, were harrowed acroſs, *ſurfaced* *, and re-harrowed. This left the Soil in gentle waves; high enough to ſhed off the ſuperfluous rain, without appearing unſightly. The re-harrowing was done by a pair of ſmall Harrows, and was intended, firſt, to loosen ſuch young Seedling-weeds as might have been preſſed into a ſtate of Vegetation by the rollers of the Land-Plane; and, in the next place, to prepare the immediate Surface for the reception of the Graſs-Seeds, and, at the ſame time, to raiſe a ſufficiency of looſe Mould for the Rakes to lay hold of in the act of covering them.

SEED-PROCESS.

As ſoon as the Oats were up, the Graſs-Seeds were ſown, and hand-raked in with common Hay-Rakes †.

The Seeds hit remarkably well; not a bald patch in either Field ‡.

VEGETIZING PROCESS.

Swept and rolled.

The Docks were intended to have been drawn; but the ground was too dry until the Graſs got too high for the operation.

* Rendered ſmooth by the LAND-PLANE, or SURFACE, deſcribed in the DIGEST of the MINUTES, page 54; and of which a Drawing is there given.

† The Oats were ſown the 29th and 31ſt of March; the Graſs-Seeds the 1ſt of May.

‡ See EXPERIMENTS, No. 64 and 65, relative to the Seed-Proceſs of Mixgraſs for 1779.

The

MIXGRASS, 1778.

VEGETABLE PROCESS.

The whole was *mown* (for the reasons already set forth), and the Docks picked out of the Swaths ; which were afterwards turned and re-picked. As soon as the Swaths were sufficiently withered, they were run into beds, with Corn-Forks ;----turned;---and carried in fine order. The ground being hot, and the Weather settled, this Hay was not cocked.

CROP.

Various : See the Account of PRODUCE at the Head of this Article.

QUONDALS.

After one plowing, the Quondals are universally foul : After Fallow, uniformly clean and promising ; excepting the upper part of Foot-path Field (T 3).

This favourite Field (See a MINUTE of 20 May, 1775.) no longer retains its former countenance : The extreme wetness of the Summer 1777, did, as I then conjectured it would, (See OBSERVATIONS 1777, page 43.) very much damage the Roots ; and the Field being mown again this year*, the puny shoots could not struggle with the *Couch* and *Watergrass*, which now almost wholly

---

* There was a reason, besides those above-mentioned, for not *pasturing* this Field : the young Quick Hedge which divides it from H and I, (and which is particularly mentioned, and referred to, in page 69 of the DIGEST) was not yet high enough for a Fence ; and the temporary *Brush* Hedge (the last I mean to make) was too much decayed any longer to answer that Purpose ; and to have raised a *new* Hedge, under the Circumstances above related, would have been doing what I judged I had no *right* to do; notwithstanding I then hoped and expected to have continued the Farm to which this Fence belongs. This circumstance is mentioned to strengthen the Apology for mowing this Field ; which, in the predicament wherein it stood, was a crime almost unpardonable.

occupy

occupy the Surface: And nothing, except pasturing three or four years successively, can save this part from the Plow and Harrow.

From the miscarriage of this once promising Ley, some useful lessons may be drawn.

IN FUTURE,

*If, through a continuance of wet weather, or a hurry of business, a Fallow, which is prone to Root-Weeds, cannot be cleansed in one Year; I will let it lie fallow another and another, rather than ley it before it be perfectly cured ** *.

I will not suffer the Herbage of a new Ley to stand until it be fully ripe †.

I will not mow a new Ley two years together, except it be remarkably clean, in extraordinary heart, and fully stocked with Plants.

* This Field is very subject to *Couch*, and the Summer of 1774 was remarkably *wet*; and although it had as good a Spring-Fallow as the hurry of the fine Spring of 1775 would permit, yet this Field remains a strong evidence that a *Spring*-Fallow cannot be *depended* upon.

† For, even in *dry* weather, the Roots may, *in some degree*, suffer.

MEADOWS.

MEADOWS.

1778.

		containing		produced		laid at		or trusses an Acre
D 2, 3.		3⅞ Acres.		8 Jags		7 Loads.		65.
L 1.		1 ——		2 ——		1¼ ——		63.
T 1, 4.		5¾ ——		7 ——		7 ——		47.
K 1, 3.		13½ ——		18 ——		15 ——		40.
P, R.		5⅛ ——		4 ——		3¾ ——		26.
		28¼ Acres.		39 Field Jags.		34½ Sale-Loads.		

S O I L.

D, L, T, K ; *clayey* Loam.

P, R ; *gravelly*, fandy Loam.

This Year, being inclinable to dry, the cool clayey Loams have greatly the advantage of the upland burning Soils.

M A N U R E.

None laid on MEADOW for this year.

S E E D.

W E A T H E R.

Inclinable to *dry*; yet very feafonable.

S U C C E S S I O N.

R 1 and K 3, after *Pafture.*

The reft after *Hay.*

K 3, which had been *hayed* for ten, perhaps twenty, years fuc-ceffively, was laft year *paftured*, in order to improve its *Herbage*; and the effect fully anfwered the expectation; the Hay of this Field,

4   which

MEADOWS, 1778.

which had ufually been harſh and *benty*, being this year remarkably full of *Leaf-Graſſes:* and generally,

Perhaps ;---*To produce a* hard, benty Hay, *frequently* mow : *to gain a* ſoft, herby Hay, *frequently* paſture.

The THEORY of this rule is at leaſt *probable*; for, if the *Blade-Graſſes* are ſuffered to ſtand for a Crop, they gain ſtrength, predominate, and ſhed their Seed ; while the *tender Herbs* are checked, if not wholly ſmothered, by the overſhading and dripping of the ſpreading Panicles of the Benty-Graſſes but, by Paſturing, the Fruit-ſtalks and Panicles of theſe become ſtinted, puny and ſtraggling; while the dwarfiſh Herbs have an opportunity of gathering ſtrength, tillering, ſpreading perhaps by being trodden, and, if not eaten too bare, ſome of them of Seeding.

SOIL-PROCESS.

SEED-PROCESS.

VEGETIZING PROCESS.
Swept and rolled.

VEGETABLE PROCESS.

Began mowing about the 8th of July, and finiſhed carrying the 27th of the ſame month.

The weather being uncommonly fine, with every appearance of its being ſettled, I beſtowed a particular attention to the making of this Hay.

The Swaths were broken, preſently after the Sithe, into thick rows; which were afterwards turned, and, in the evening, gene-

3   rally

rally made into fmall Cocks, proportionable to the degrees of graf-
finefs of the Herbage.

Next day thefe Cocks were broken into Beds ;—carefully turned;
and carried to the ftack almoft as green as the Grafs was before it
was mown *.

Thefe feveral operations were intended to mix and affimilate the
Hay, without expofing it too much to the intenfe heat of the Sun.
No unbroken tuft, nor green lock, could efcape the feveral break-
ings and turnings ; each part had the benefit of the Sun, without
being expofed to it too long ; confequently, every part became
equally, deliberately, and thoroughly made.

The motive for *cocking* this Hay was two-fold : the firft, one
day's Sun was not fufficient to make it, and it was cocked in the
evening to guard againft a fquall in the night : the other motive
was founded on the following THEORY.

The Herbage of a Meadow is compofed of a variety of Plants,
which mature at different periods. If a Meadow be fuffered to ftand
until its *Bottom-Graffes* have reached their perfection, the *Bents* † are
of courfe become ftrawy and faplefs.

Before the Herbage of a Meadow can be ftored up as Winter-
fodder, it muft have diffipated a principal part of its fap or juices.
If Grafs be *thinly* expofed to the fun and air, thefe juices are of
courfe immediately abforbed by the Atmofphere; and, if expofed too
long, its nutritive quality may be confiderably injured. On the
contrary, if the Herbage be kept in *thick* Rows, or Beds, and more

* The ground being hot, and the weather parching, I was afraid of *making it too
much*, and thereby preventing its fermenting properly in the Stack : This precaution
was, *perhaps*, too rigidly obferved; for the Stack being large, the middle of it cut
out *fomewhat brown*.

† Perhaps, a provincial Term : The *Straw* ;—the *Stems*, the *Fruit-ftalks* of the
Blade-Graffes.

efpecially

especially if it be *incorporated* in a Heap or *Cock* of any size; is it not rational to suppose that the juicy Grasses, which are in the very act of dissipating their superfluous Sap, will communicate at least their Flavour, if not some share of their nutritive quality, to the parts which are dry and sapless? It is not at least rational, that many essential Particles will be retained, which, had the Hay been more *thinly* exposed, would have been communicated to the Atmosphere?

PEASE.

PEASE.
1778.

H 1, 2.	containing	10 Acres.	produced	22 Jags.	laid at	30 Quarters.	or bushel an-Acre.	24.
P 3.		4 —		9 —		12 —		24.
P 1, 2.		3 —		6 —		8 —		21.
		17 Acres.		37 Field Jags.		50 Quarters.		

S O I L.

H, a *sandy* Loam, with a cold, *springy* subsoil.

P 3, a *clayey* Loam intermixed with *Gravel*, with a *retentive* Subfoil; which, however, is tolerably *found*.

P 1, 2, *gravelly* Loam, with an *absorbent* subsoil.

The Pease of H, (white boiling Pease) were sown the 12th of March, and *came up* remarkably well;—the Plants were numerous, even, and healthy; but, during April and May, many of them *went off*, dwindling away imperceptibly; insomuch that a patch or two, which lie *wetter* and *colder* than the rest, were left entirely destitute of Plants. About a quarter of an Acre was sub-plowed,—harrowed,—the trumpery raked out,—and the patch sown with turnips. I had almost determined to have served a principal part of the Ten Acres in this manner: it was lucky, however, that I did not; for, when the Subsoil became *less chilling*, and the Surface had been moistened by some *warm showers*, the surviving Plants shot up with uncommon rapidity; the halm on the thin patches ran to six or eight feet long; and even on the *rottenest* parts the Crop was not bad: where the Soil lies tolerably dry and comfortable, the Crop was very good. On the whole, *as the Weather happened*, the Crop was moderately good: But I am firmly of *opinion*, that had the Spring proved *cold* and the Summer *wet*, the Crop on the whole would have scarcely been worth harvesting. Therefore,

P

Perhaps;

PEASE, 1778.

Perhaps ;---*Let the* Soil *be ever so eligible, if the* Subsoil *be* unsound, *it is at least* hazardous *to sow* white Pease.

Those of P 1, 2, (Marlborough Grays) were sown very early (the 12th of February) ; but the weather setting-in frosty and cold, they did not come up until near the middle of March ; from which time, the *Subsoil* being *absorbent,* they never received the smallest check.

P 3, (likewise Marlborough Grays) was sown the 14th of March ; these received a check about the same time when those in H went off ; but the Subsoil, though retentive, being *sound,* the roots of the Plants remained untainted ; and as soon as they had received some warm showers they came away apace, and gave a *good Crop,* not-withstanding the *Soil* is much *stiffer* than what is generally esteemed a *Pea-soil.* Therefore,

Perhaps ;---Gray Pease *may be eligibly sown on a* stiffish *Soil* with a retentive *Subsoil, provided the Subsoil be* found.

MANURE.

H, *well manured* with *compost, laid on* for the preceding Crop.

P 2, 3, *manured* with *dung, plowed in* for the preceding Crop.

P 1, *well manured* with *Compost, laid on* for Crop 1775.

SEED.

H, *White* boiling Pease, bought at the London market. (*Soil* unknown.)

P, Marlborough *Gray* Pease, immediately from Wiltshire. (*Soil* likewise unknown.)

WEATHER.

From the beginning of Harvest 1777, to the end of Harvest 1778, we had not, generally speaking, any rain. The *Winter* was remarkably free from wet, and so were *Spring* and *Summer* ; excepting a few genial showers before and after Hay-time.

It

It feems to be a generally received opinion, that Peafe and other Pulfe affect wet weather: the Crops of this year, however, prove that opinion to be erroneous, for fuch a *Pea-Year* has not lately happened; the halm is not only luxuriant, but it is remarkably well podded.

Therefore;—*Peafe require* Warmth and Moifture, *rather than* Wet.

SUCCESSION.

H, after *Clover*, after *Wheat*.

This Ley would have been too foul for Wheat on one plowing; and, in regular fucceffion, it was to be Wheat in 1779. It was therefore fowed with Peafe, in order to receive the benefit of a Dog-days Fallow: and, had this intention been profecuted, it would have afforded an excellent Wheat Fallow.

P 2, and part of P 3, after *Wheat*, after *Fallow*.

Part of P 3, after *Wheat*, after *Clover*.

P 1, after *Wheat*, after *Rye-grafs* and a Dog-days Fallow.

The Crops were moderately *good* in every inftance.

SOIL-PROCESS.

The whole once plowed with the *Buryfod Swing-Plow*; excepting part of P 3, which was too *woclly* for the Buryfod to act *: This part, however, having been Summer-fallowed for the preceding Crop of Wheat, the operation was not *neceffary*.

In H, the Buryfod made exceedingly good work: the Beds, after they had been harrowed, looked rather like *Garden* than *Field-Beds*;

* By grinding the Edges of the Share and Coulter-parts very fharp, they might have operated; but this part being very clean, and in exceedingly high Tilth, there was in fact no *fod*, nor fcarcely any Weeds, to bury. The Buryfod Plow operates the beft where the Surface is firm.

and

PEASE, 1778.

and the Pea-Quondal is much lefs *couchy* than the Clover-Ley was before plowing. (See SUCCESSION.)

Therefore ;---*Burying the Sod of a Clover-Ley for Peafe is eligible management*.

SEED-PROCESS.

The whole fown *broad-caft* over the *frefh Plit*. For the *Times of Sowing* fee SOIL.

VEGETIZING PROCESS.

Part difweeded ; but the Peafe grew fo quick and luxuriantly, the reft got too high to draw the Weeds without injuring the Crop.

VEGETABLE PROCESS.

Principally cut with Sithes.

Part of them were mown by the day, part by the Acre.

It is difficult to get labourers by the Acre to take fufficient pains in making them into proper *Wads*, *Reaps**, or *Bundles :* they are apt to make them too *large*, and too *flat* and *ragged* ; leaving them fcattered in the Interfurrows, or wherever they happen to be made. Whereas a Wad of Peafe fhould, in fome meafure, refemble a globe or *ball* of two feet to two and half feet diameter, and be placed in

---

* In the North of England they are called *Reaps* ;. and fo are the *unbound Parcels* of *Wheat*, or other Grain.   And perhaps *Reaping*, as applied in the South of England, where it means *Cutting*, is a corruption of this term. In the Northern parts of the Kingdom (where the veftiges of the Anglo-Saxon language are ftill very ftrong) this Term is not ufed ; except when applied to *Peafe*.   The *Cutting* of *Wheat* is termed *Shearing* ; probably from the inftrument with which our forefathers cut down their fcanty patches of that ineftimable Grain. *Cutting*, or *Cutting-down*, feems to be the moft general, and the leaft ambiguous Term.

I   *rows*

PEASE, 1778.

rows (or *a-zig-zag*, if one row will not contain them) on the *ridges* of the Lands, (fuppofing them to be half-rod Lands) or, if the Lands run eaft-and-weft, on the *fouth-fide* of the ridges.

This operation, however, is difficult and tedious to be done with a Sithe and the Foot ; it is therefore more eligible to perform it with a Prong ; requiring of the Mower (whether by the Day or the Acre) no more than making the Reaps of a proper *Size*, without his attending either to the *forming* or the *placing* of them.

In dry weather, Peafe properly wadded with a Prong are much fooner ready to carry than thofe left in hard bundles by the Foot and Sithe *, without form or method ; and in cafe of wet Weather, they are not fo liable to open, nor turn black ; their contact with the earth being fmall, and that, too, with the drieft part of it.

CROP.

On the whole good. See the ACCOUNT OF PRODUCE at the Head of this Article.

QUONDALS.

In general cleaner than could have been expected from the foulnefs of fome of the Soils before plowing, efpecially the Clover-Ley of H.

---

* Some men, it is true, will fet them up very light and *hollow* with the foot ; but thefe are rarely to be met with ; nor will a *hollow* Reap bear turning fo well as a *porous* Reap ; and a man may be taught in five minutes to fet them up high with a Prong. It is done by putting the Prong into the middle of the Wad, and, by a fudden fhake, rendering the inner parts cellular, or porous, and forcing down the ragged outer fkirts ; which, by a circular motion of the hand, are formed into a kind of foot for the Reaps to ftand upon. The Prong fhould be put in and drawn out perpendicularly, with the right hand placed clofe to the Tines.

I will

PEASE, 1778.

IN FUTURE,

I will never fow white *Peafe over a* fpringy *Subfoil* †.

I will venture gray *Peafe on* ftiff *land, provided the Soil be in* Tilth, *and the Subfoil* found.

I will ever depofit a retentive *Soil for Peafe in* high *half-rod Ridges.*

I will endeavour to bury the Surface *for Peafe ; leaving the Seed-Seams as* open *and* deep *as poffible.*

I will not fow Peafe early *over a* retentive *Soil,*

I will fow a retentive *Soil* broadcaft *over the* frefh Plit.

I will fow a light abforbent *Soil as* early *as it can be* drilled *or* fluted *with propriety.*

I will be careful to hand-weed *the random-fown, and to* hoe *thofe fown in* Rows, *before the Weeds get too high.*

I will endeavour to cut *Peafe with* Sithes, *before they be too* ripe, *and* reap *them with* Prongs.

† Notwithftanding the Seed fown in H was a beautiful fample, and an uncommonly fine boiling Pea, the produce is unfightly, and boil very badly.

WHEAT.

W H E A T.
1778.

S.		containing	1½ Acres.	produced *	3 Jags.	laid at	5 Quarters.	or bushels an-Acre.	26½
I 2			6¼ ——		9¼ ——		20 ——		24¼
O.			11 ——		18½ ——		32 ——		23
N.			8 ——		10½ ——		22 ——		22
I 1.			5½ ——		6 ——		12 ——		17½
			32¼ Acres.		47¼ Field-Jags.		91 Quarters †.		

SOIL.

S, *sandy* Loam, with a springy Subfoil.

I 2, stronger Loam, with a retentive Subfoil

O, the same, with the same.

N, gravelly Loam, with a varied Subfoil.

I 1, *sandy* Loam, with a springy Subfoil.

Thus it appears that. *sandy Loam* gives the best, and also the *worst* Crop; and what renders the *Quality* of the Soil of still less importance to Wheat, the Soil of S 3, is of an inferior quality, and is worse situated, than that of I 1.

Therefore;—*The Quantity of the Crop of Wheat had no dependance on the Quality of the Soil.*

MANURE.

S, 600 feet in LONG-DUNG, *plowed in* for 1777.

I 2, 600 —— COMPOST, *laid on* for 1777.

O, 500 —— SHORT-DUNG, *plowed in* for 1778.

N, 400 —— VARIOUS, *plowed in* for 1778.

I 1, 500 —— COMPOST, *laid on* for 1777.

* Began *carrying* the 10th, and finished the 19th of August.

† This Estimate is too high: it was made as the Corn was carried. Wheat this year does not yield so well as might have been expected. (See WEATHER.)

Therefore;

Therefore;—*This Year's Crop of Wheat was nearly in Proportion to the Quantity of Dung.*

Excepting;---*That* 600 *foot in Long-dung, plowed in, was preferable to* 600 *foot in Compost, laid on.* But,

By Experiment, No. 28;—*The Compost laid on* I 2, *was of no obvious service to this Crop.*

Therefore;---*Nothing conclusive, with respect to Manure, can be drawn from this Year's Crop of Wheat.*

How abstruse is this department of Agriculture!

S E E D.

The whole *white Wheat* raised last year on Soils opposite to those on which it was sown this year. The CHANGE was from *clayey Loam* to *gravelly Loam*; and from *gravelly Loam* to *Loam* and *sandy Loam*. See the ARRANGEMENT.

No comparative Experiment was made.

W E A T H E R.

The whole year remarkably free from Rain: yet the ground was not *droughty* until late in the Summer; when it became as impenetrable as a pavement.

Perhaps a year so uniformly indulgent to the Farmer has scarcely ever happened.—The *Autumn* was fine to sow his Wheat in.—The *Winter* was temperate.—The *Spring* dry, to get in his Spring-Crops. Some fine rains between Spring Seed-time and Hay-time.—Hay-time uncommonly fine.—Between this and Harvest, a charming rain: And the Harvest uniformly fair! Nothing seemed to be wanted, which even a Farmer could have wished for excepting a few moderate showers immediately before and during Harvest, to plump the Grain in the ear; and perhaps the backward-sown Wheats were some of them injured by a thunder-storm, which happened just when they were blooming. But, taken all-in-all, another

year

year fo favourable to the Farmer may not happen for an age to come.

It is generally underftood that a dry year is good for Wheat. This year and its Crops corroborate that opinion.—The Burden is large, and the Ears long; yet Wheat this year *yields* lefs than was expected; owing probably to the two unfavourable circumftances abovementioned: namely, the Fecundation being interrupted, and the droughtinefs of Harveft.

SUCCESSION.

S 3, after *Potatoes*, after *Turnips*.

I 2, after *Clover*, after *Wheat*, after *Wheat*, &c.

O, after *Clover*, after *Clover*, after *Barley*.

N, after *Clover*, after *Barley*, after *Tares*, &c.

I 1, after *Clover*, after *Barley*, after *Wheat*.

Potatoes, *Wheat*, gave the beft Crop, owing probably to the fu- perior Heart and Tilth of S 3.

Clover, *Wheat*, was good or bad, nearly in proportion to the *cleannefs* of the refpective Fields. I 2, which is tolerably *clean*, gives 24½; while I 1, (a nearly fimilar Soil, with nearly fimilar Manage- ment) which is *foul*, gives only 17½.

Therefore ;---*Foul Clover-Leys are unfit for Wheat.*

It is obfervable that part of I 2 has born three Crops of Wheat in four years: and, what is ftill more worthy of notice, the two latter were much the beft Crops.

This patch was Beans in rows, in 1774; and the Soil being tolerably clean, the Weeds were kept under by repeated hoeings; and after the Beans were *drawn*, the Soil was fubplowed, and landed- up for Wheat. The Crop very indifferent. The adjoining part of the fame Field was likewife Beans in rows; but the Soil being foul, the Hoes were unable to fubdue the Weeds; and, inftead

Q

of

WHEAT, 1778.

of Wheat, this Part was Summer-fallow in 1775. In order to bring the firſt-mentioned patch into the ſame Crop in 1776, the Wheat was mowed, (See a MINUTE of 7 Auguſt, 1775) the Quondal was plowed three or four times, had a ſprinkling of Dung, and was landed-up with the reſt of the Field for Wheat. The Crop very good. In the Spring of 1776, Clover-Seed was ſown over the Wheat; and, in the Winter 1776-77, the whole was very well manured with Compoſt laid on the young Seeds: and, in Autumn 1777, the Clover Ley was broken-up for the Crop of Wheat under conſideration. The Crop good. Therefore,

Perhaps;—*The* Succeſſion *is leſs eſſential to Wheat than are* Manure *and* Tillage.

It is evident, however, from the Crops of **N, O**, that the *Succeſſion* of CLOVER, WHEAT, is *not ineligible* on a ſandy, gravelly Loam, the Crops being *uncommonly good* for the Soil they grew on; eſpecially where the Dung was *plowed-in.----Potatces*, or even a *dunged Summer-Fallow*, would probably not have given one ſo good; and the *Conveniency* of Clover, Wheat, is ſtrikingly obvious.

Therefore;---CLOVER, WHEAT, *is a moſt eligible Succeſſion for a ſandy, gravelly Loam.*

SOIL-PROCESS.

S 3, plowed three times with intermediate harrowings.

The Clover-Leys, plowed once with the *Buryſod Swing-Plow.*

Some of the Leys, eſpecially thoſe of two years old in O, were very foul and graſſy; and had they been plowed in a ſlovenly manner, the *Manes* of the Plits would, before Harveſt, have matted the Surface, the Crop would of courſe have been injured, and the Quondals left as green as Leys; whereas, by burying the ſoddy edge of the Plit, they are now *lefs grafsy* than the Leys were before plowing. The additional expence is trifling : it cannot be more than half a horfe, or about nine-pence an Acre.

Therefore ;—*Burying the Sod of a Clover-Ley for Wheat is moft eligible management.*

It is true, that a *good*, or rather an *excellent* plowman will plow *flat Beds* in fuch a manner as to *tuck-in*, or *bury*, the Sod without cutting it off; but the beft plowman cannot poffibly *reverfe round ridges* in any way which will nearly anfwer the purpofe of cutting off the foddy edge of the Plit, and burying it in the bottom of the preceding plow-furrow. And when we confider how many *indifferent* plowmen there are to one who is *excellent*, a BURYSOD PLOW is a great acquifition to Tillage; and if we reflect how much land there is in this Ifland which cannot with any degree of propriety be plowed *flat*, and, of courfe, cannot be plowed properly with a *Wheel*-Plow, a Buryfod *Swing*-Plow has an indubitable Claim to the attention of every *wet* land Farmer *.

SEED-PROCESS.

Part of the Seed was *prepared*.

Part was fown *dry*.

By Experiment, No. 61 ;—*Neither benefit nor difadvantage arofe from* pickling *the Seed of Wheat*.

The whole was fown *broadcaft* over the *frefh Plit*, in *October* and *November*.

The *Quantity of Seed*, on a par, 2½ Bufhels an-Acre.

VEGETIZING PROCESS.

In general hand-weeded.

* This (I hope) is not fpoken through an overweening Partiality ; but proceeds fpontaneoufly, from a thorough conviction of the truth of this affertion.

VEGE-

WHEAT, 1778.

VEGETABLE PROCESS.

The whole cut by the Acre; at ten to thirteen Shillings an Acre.

CROP.

See the Head of the Article.

QUONDALS.

S 3, very clean.

The reſt in proportion to the Cleanneſs or Foulneſs of the Leys before plowing. See the SUCCESSION.

GENERAL OBSERVATION.

I have not from this year's management found any reaſon to alter the reſolutions I formed laſt year with reſpeċt to the Wheat Proceſs.: and I can only add this year,

IN FUTURE,

I will not, be my Plowman ever ſo excellent, plow a Clover-Ley for Wheat without burying the Sod.

OATS.

O A T S.

1778.

K 4. $\Big\}$ containing $\Big\{$ 6⅜ Acres. $\Big\}$ produced $\Big\{$ 11 Jags. $\Big\}$ laid at $\Big\{$ 30 Quarters. $\Big\}$ or bush. an Acre. $\Big\{$ 38.

L. $\Big\}$ $\Big\{$ 24 —— $\Big\}$ $\Big\{$ 47 —— $\Big\}$ $\Big\{$ 90 —— $\Big\}$ $\Big\{$ 30.

30⅜ Acres. 58 Field-Jags. 120 Quarters.

S O I L.

The whole *clayey Loam*, on *retentive* Subfoil. Neverthelefs **L** **1** and 2, have always been efteemed *good* Land, and K 4, the contrary; fo much fo as to obtain the Name of *Small-Profit*.

M A N U R E.

In order that K 4, might no longer deferve the ftigma it had received, this Field was Summer-fallowed, and *well-dunged* for Oats and Ley-graffes.

L 1, 2, were *well-dunged* for 1777.

S E E D.

The Seed was various; chiefly Scotch Oats raifed on a chalky Loam: but the untowardnefs of circumftances prevented any Experiment, or accurate Obfervation from being made.

W E A T H E R.

See WHEAT, page 96.

S U C C E S S I O N.

K 4, after *Summer-Fallow*, after *Wheat*.

L, after *Wheat*, after *Fallow* or *Fallow-Crops*.

By the terms of the Leafe, L 1, 2, were to be leyed in 1779. Confequently they *ought* to have been Fallow-Crop in 1778, for Oats and Mixgrafs in 1779.

The

O A T S, 1778.

The Crops of both were *large*; but the *Summer-Fallow yielded* much the beft, and gave, beyond Comparifon, the fineft *Sample*.

Soil Process.

K 4, plowed fix or feven times, with intermediate harrowings, &c. the Wheat-Quondal, from which the Fallow was made, being exceedingly foul.

L, was once plowed with the Buryfod Swing-Plow; and left a Quondal much fuperior to what could have been expected from *Wheat, Oats*.

Seed-Process.

Part of L 1, was fufficiently harrowed.

Part (through an unavoidable want of felf-attendance) was badly harrowed; the Plits being left almoft unbroken, and the Seed confequently expofed above ground.

The former produced a very good Crop; the latter, a thin, ftraggling Crop, and that almoft fmothered with Weeds. The difparity arifing wholly from the difference of *Harrowing*, was not lefs than two guineas an Acre!

Therefore;—*A fhilling faved in* covering *was a Guinea loft in the Crop.*

The *time of Sowing* was from the middle of March to the latter end of April.

K 4, being in very high Tilth and Heart, I was afraid the Crop would be too rank, and, by lodging, would fmother the young graffes; I therefore deferred fowing this Field until the middle, or towards the wane, of April. I endeavoured to regulate *the Time of Sowing*, and *the Quantity of Seed*, to *the State of the Soil*; fo as to have a full Crop of Oats, without injuring the Ley-Graffes: and I was fortunate enough in my Regulation; for,

By

By Incident in K 4 ;---*Four Bushels of Oats sown in the wane of April over a well-dunged Summer-Fallow, gave a* large, *yet a* standing *Crop.*

This Incident serves to remove an objection to *fallowing* and (if the Soil be much out of heart) *dunging for Spring Corn and Clover* or *Ley-Grasses* ; as by hastening or protracting *the time of Sewing*, and by encreasing or diminishing *the quantity of Seed,* a Farmer may (the uncertainty of the Weather apart) have almost *any* Crop of Spring-Corn he chuses. Consequently the *Dangers of Rankness* may be at any time avoided ; especially of *Oats.*

VEGETIZING PROCESS.

The sowing of the Grass-Seeds of K 4, was unavoidably postponed until the Oats had got to a considerable height ; and the Surface of the Soil was of course become stale. In order to raise fresh Mould for the Seeds to fall upon, the whole Field was run over with a pair of light Harrows ; excepting a belt acrofs the middle, left experimentally unharrowed.

By Experiment, No. 66 ;—*Oats, when six or eight inches high, may be run over with a light Harrow, without injury.*

VEGETABLE PROCESS.

Mown while the knots in general were still green. Lay in Swath until the Sap was wholly dissipated ; and carried with very little lofs by shedding.

THE CROP.

On the whole very good. Part of L 1, was exceedingly *bulky* ; but the *Grain* is *thin.*

The *Sample* from the *Summer-Fallow* is at least ten per Cent. superior to that from the *Wheat-Quondal.*

QUONDAL

OATS, 1778.

QUONDAL.

K 4, perfectly clean; except a part which was caught in the wet of the Summer of 1777, and which got so grassy during the continuance of the Rains, that it never was thoroughly cured.

L, is much cleaner than could have been expected after two *Corn-*Crops immediately following each other; the part Summer fallowed especially. The distinction between this part and the part fallow-cropped is *obvious* at the distance of at least *two Miles*. The difference does not arise so much from *Couch*, or *Grass*, as from *diminutive Seed Weeds*; from which a *whole* year's Fallow purges the vegetable stratum; but to the destruction of which a *partial* Fallow is inadequate.

GENERAL OBSERVATION.

The result of this year's management of Oats serves to strengthen the resolutions I made last year; which see in page 68.

BARLEY.

BARLEY.
1778.

M.	containing	18¾ Acres	produced	38 Jags	laid at	65 Quar.	or Bushels an Acre.	28.
S.		1½ —		3 —		5 —		27.
F.		9½ —		28 —		32 —		26.
		29¾ Acres.		69 Field-Jags.		102 Quar.		

S O I L.

M, part *Loam* with a *retentive*, and part *gravelly Loam*, with an *abforbent* Subfoil *.

S, a *fpongey Loam*, with a *fpringy* Subfoil.

F, the fame, with the fame.

M A N U R E.

M, *varioufly manured* for 1775 and 1776.

S 1, *well dunged* for 1777.

F, *dunged* for the prefent Crop of Barley.

Thus it appears that the *Yield* is in direct oppofition to the *year of manuring*. It muft be obferved, however, that the *Burden* bears an affinity to the MANURE; and the contradiction in the *Yield* arifes from the SOIL; the Soil of S, and F, wanting perhaps a fufficient *Texture* to produce a *yielding* Crop.

S E E D

M, raifed on a *chalky* Loam.

F, S, on a *gravelly* Loam.

W E A T H E R,

See WHEAT, page 96.

* Their feparate Products were not noticed at Harveft.

R

S u c-

BARLEY, 1778.

SUCCESSION.

M, after *Fallow-Crops*.

S, after *Cabbages*.

F 1, after *Summer-Fallow*.

F 2, after *Tare-Barley*.

Part of M was after *Peabeans*.

Part after *Tare-Barley cut for Hay*.

Part after *Tare-Barley cut for Verdage*.

Part after *Tare-Barley plowed-in*.

By Experiments, No. 48, 49 ;---*It feems to be more eligible to* verdage *than to* bury *a foul Crop of Tare-Barley*.

SOIL-PROCESS.

The *Summer-Fallow* was, in Autumn 1776, laid-up into Ridglits; part of which were experimentally crofs-plowed in Winter, part left entire until after Spring Seed-time. The latter was much the beft Fallow. (See a MINUTE of the 18th of June 1777.)

The Parts *Fallow-Cropped* were *trenched* (thrown up into Ridglits) as foon as the Crops were off; and having laid fome Weeks in this expofed ftate, they were crofs-plowed before the hurry of Wheat Seed-time came on. After the Wheat-Seed was in the ground, they were thoroughly harrowed, and prefently afterwards *re-trenched* and crofs-furrowed for the Winter *. The beginning of April, the Peafe and the principal part of the Oats being in, the Winter Ridglits were thrown down by another crofs-plowing ;—harrowed ;--comprefled

---

* This was the firft time I had trenched a *Fallow :* I had frequently laid a *firm Soil* into Ridglits, by laying an entire Plit upon an unftirred Balk; but had not raifed *loofe Mould* into a fimilar ftate, by reaching about twenty inches with a wide fterned Plow; which not only leaves the Mould in regular Ridges, with a *neatnefs* beyond the art of the Gardener to excel, but the whole Soil is more-or-lefs *ftirred*.

3

very

very hard with the Land-plane;—harrowed;—and, the remaining Weeds having laid a few days to wither, gathered up into five-bout Beds for the Seed.

The Crop was uniformly large: but why it fhould be fo on Land which had been exhaufted by three fucceeding Crops, cannot eafily be accounted for; except by its being *expofed by Trenching:* firft to the fcorching heat of Dog-days, and afterwards to the frofts of Winter. And fcarcely any Premium would induce me to break-up a Fallow, or to expofe it, during Winter, in any other way, than by raifing it into Ridglits with *deep* Trenches between them *.

S E E D - P R O C E S S.

The whole fown *broadcaft* over the *frefh Plit.* The *Quantity of Seed* fomething more than two Bufhels an-Acre.

V E G E T I Z I N G   P R O C E S S.

F 1, 2, remained unrolled (in hopes that it might have become the intereft of fomebody to have fown them with Clover) until the Barley was beginning to fpindle.

By Experiments, No. 62, 63;---*Barley may be rolled when it is fix or eight inches high, without much injury to the Plants.*

V E G E T A B L E   P R O C E S S.

It was fuffered to ftand until the ears began to *curl;*—lay in Swath until the remaining juices were exhaled; and carried, as much as poffible, during the heat of the day.

* A Fallow which requires *Spring-ftirrings,* is here fpoken of. If a Fallow (a *wet-*land Fallow at leaft) be perfectly cured, fo as to require only *one Spring plowing,* it fhould, in Autumn, be gathered in half-rod Ridges, in order that it may, in the Spring, be reverfed for Clover, or thrown two-into-one for Ley-Graffes.

C R O P.

BARLEY, 1778.

CROP.

See the Head of the Article.

QUONDAL.

Uniformly *clean*; yet destitute of *Clover!* How provoking! Thirty Acres of Land which had engaged my particular attention during *two* years, and which were the first I had managed, (on a large scale) in a manner which I cannot forbear wishing to recommend to others, were, by the jarring of parties, to fall below the level of ordinary Management, for want of the finishing operation. Instead of *Spring-Corn, Clover*, the principal part of these thirty Acres have been subjected to the *barbarous* Succession of *Barley, Wheat!*

But I will not dwell longer on a subject which is painful to me, and which nothing but an earnest desire of promoting, to the utmost of my information, the advancement of Agriculture, could have induced me to have undertaken; much less to have entered into so minute a detail.

PRODUCE 1778.

$13\frac{1}{4}$	Acres of CLOVER,	produced $14\frac{1}{2}$	Jags, laid at	13 Loads.
$49\frac{1}{8}$	——— MIXGRASS,	——— $51\frac{1}{2}$	——, ———	44 Loads.
$28\frac{3}{4}$	——— MEADOW,	——— 39	——, ———	$34\frac{1}{2}$ ———
17	——— PEASE,	——— 37	——, ———	50 Quarters.
$32\frac{1}{2}$	——— WHEAT,	——— $47\frac{1}{2}$	——, ———	91 Quarters.
$30\frac{3}{8}$	——— OATS,	——— 58	——, ———	120 ———
$29\frac{3}{4}$	——— BARLEY,	——— 69	——, ———	102 ———
$200\frac{3}{4}$	Acres.	$316\frac{1}{2}$ middling Field-Jags.		
$4\frac{5}{8}$	——— PASTURAGE,			
$4\frac{1}{2}$	——— VERDAGE,			
$27\frac{1}{4}$	——— FALLOW,			
32	——— GIVEN-UP, &c.			
$21\frac{7}{8}$	——— HEDGES, &c.			
291	Acres.			

HAVING

HAVING thus given a *syftematic View* of the Tranfactions or Management of 1777 and 1778, fo far as relates to the principal objects of Farming, namely, AGRICULTURAL VEGETABLES, I fhall now proceed to a more *general Review* of the Occurrences of thefe years *.

The Articles already noticed are—*Wheat,—Barley,—Oats,—Peafe, Tares,—Peabeans,—Tare-Barley,—Meadow,—Mixgrafs,—Clover.*

The Heads which remain to be treated of are, *Farms,—Soils.— Manure,—Seed,—Weather,—Servants,—Beafts of Labour—Implements,—* and the feveral other Heads enumerated and difcuffed in the DIGEST of the MINUTES OF AGRICULTURE.

In taking this Review, I fhall firft offer fuch information as may have prefented itfelf fince the publication of that work, and which is efpecially applicable to the fubject or Head under confideration ; and afterwards *clafs*, or bring into one point of view, fuch Experiments and Obfervations, already enumerated, as have a joint reference to the Article or Head in review.

F A R M S.

NO frefh information relative to this fubject having arifen fince the period abovementioned ; and as it is not my intention to recapitulate here what I have already publifhed, I will beg leave to refer to the Analyfis and Obfervations relative to FARMS, given in the DIGEST of the MINUTES, page 17.

For a Defcription of the FARM under confideration, fee the INTRODUCTION TO THE EXPERIMENTS. And for a Defcription of its FARMERY, fee the DIGEST of the MINUTES, page 21.

* The reafon for uniting the Two Years Management with refpect to thefe Mifcellaneous Articles, are given in page 70.

SOILS.

S O I L S.

EXPERIENCE convinces me, more-and-more, that the value of Land does not depend more on the Soil, or vegetative Stratum, than it does on the Subsoil, or the *Bed* of the plant-feeding Mould.

By Obſervation, page 16 ;—Perhaps every ſpecies of Loam is affected by *Clover*. But perhaps the ſtronger the Loam, the better is affected by that Vegetable.

By Obſervation, p. 34 ;---Gravelly Loam is *not* affected by *Beans:*

By Obſervation, p. 45 ;----Perhaps almoſt every Species of Soil is affected by *Wheat*.

By Obſervation, p. 65 ;---*Oats*, too, affect almoſt every Species of Soil.

By Obſervation, p. 75 ;----Perhaps tenacious clayey Loam is peculiarly affected by *Clover*.

By Experiment, No. 45 ;----Tenacious clayey Loam is better affected by *Cow-Graſs*, than it is by *finer Graſſes*.

By Obſervation, p. 85 ;----Clayey Loams are more eligible for *Meadow*, than gravelly Loams are.

By Obſervation, p. 90 ;---An unsound Subsoil is dangerous to *white Peaſe*.

By Obſervation, p. 90---*Gray Peaſe* may be eligibly ſown on stiffish Soils, provided the Subsoil is *ſound*.

MANURE.

M A N U R E.

WHILE the VEGETABLE ECONOMY remains myfterious, the THEORY OF MELIORATION will of courfe be hypothetical. EXPERIENCE, it is true, may do much; but, unaffifted by fome *general Principles*, its advances to SCIENTIFIC SYSTEM will be flow. Indeed, fo little underftood is this department of Agriculture, that even an ordinary communication of ideas refpecting it is difficult. We may fay that fuch a Field was *dunged*, or had a *dunging*, in fuch a year; but until we know that *Dung* is the FOOD OF VEGETABLES, and how much of fuch food is confidered as a *dunging*, the idea muft be entirely inadequate. It is true, a certain number of *Loads* may be mentioned; but until the vegetable Food contained in a Load be afcertained, the idea will ftill remain vague.

Whether the Vegetable Economy is or is not a myftery, will not here be difcuffed. That *Dung* * is *a* FOOD OF VEGETABLES will be *affumed*, and the enquiry will be confined to the QUANTITY OF DUNG.

In September 1777 I had an opportunity of meafuring a quantity of Dung which had lain fome time in a deep regular pit, and

---

* This Term is not made ufe of either for its elegance, or becaufe it conveys a precifely determinate idea; but for its popularity, and becaufe it conveys a more definite idea than any other term in ufe. The idea here meant to be communicated by it is not *Animal Excrements*, alone, but *any Solution of Animal or Vegetable Subftance*, whether the Solution be produced in the Animal Vifcera, or by fimple fermentation, and putrefaction. Nay, while it is yet in the act of putrefaction, or even fermentation, I have (agreeable to the common acceptation) termed it *Dung*; and the ordinary diftinction of *Long-Dung* and *Short* or *Spit-Dung*, have likewife been obferved.

A more *polite*, and indeed a more *fcientific* (becaufe lefs ambiguous) term might, no doubt, be brought forth by invention: But who is to fuffer the *Pains* of Innovation?

which

MANURE,

which was thoroughly digefted. After having moved away what lay above the furface of the pit, I levelled the top of that which remained, fo as to reduce it to *ftricken meafure*; and kept an exact account of the Loads which were afterwards carried out of the pit; which, being emptied, I took its exact gage.

The number of Loads carried were Thirty.

The gage of the Pit 2642 cubical feet.

Therefore the Carts took away 88 cubical feet of Dung each Load.

But the Carts were uncommonly large, one of them *gaging* 72, the other 85 cubical feet; on a medium $78\frac{1}{2}$ feet, which is 10 feet each Load lefs than the number carried : the *Roof* therefore contained 10 feet, befides allowing for the *Porofity* of the Loads compared with the *Compactnefs* of the Pit.

On gaging different Carts, I found that one of a medial fize did not contain 50 feet; I therefore fixed the *medial* fize of a *Load* of Dung to 50 cubical feet; and 10 of fuch Loads, or 500 cubical feet of perfectly refolved animal or vegetable fubftance, an-Acre, I confidered as a *medial dunging*. See a Note, page 30.

Or inftead of the *Number of cubical Feet*, the QUANTITY OF DUNG might be communicated by the *thicknefs of the fheet or covering*. Five hundred cubical feet of matter will cover an Acre of Soil nearly one-feventh of an inch thick ; 363 feet, exactly one-tenth of an inch in thicknefs.

It muft be obferved, that a *perfect Solution* of animal or vegetable Subftance (with moifture fufficient to make it confiftent) is here fpoken of; and, confequently, if the matter has not reached that ftate, or if it is compounded with *foffil*-manure, &c. the quantity of perfect Solution muft be *eftimated*.

It may with great truth be obferved, that, after all, the QUANTITY OF DUNG is not reduced to *Certainty*. *Certainty*, at prefent,

I

muft

muſt be left out of the queſtion; which is, whether this mode of aſcertaining and communicating the QUANTITY OF DUNG is, or is not, *more
certain* than any other which has hitherto been thought of *.

My primary intentions were to have treated this Subjeᶜt in ſome degree *ſyſtematically*; but I find myſelf at preſent inadequate to the taſk,
and ſhall therefore cloſe this Article by enumerating, miſcellaneouſly,
the EXPERIMENTS and OBSERVATIONS (including INCIDENTS and IN-
FERENCES) which have occurred during the laſt two Years managemenʳ,
whether they appertain to MANURE; MANURING; — TILLAGE; —
or HERBACEOUS MELLIORATION. (See Note, p. 51.)

By *Experiment, No.* 17;—DUNG and COMPOST were equally unſer-
viceable as *Top-dreſſings* for *Wheat.*

By *Experiment, No.* 21;—POND-DREGS *laid on* a *clayey Meadow,* in
November, were of no obvious ſervice.

By *Experiment, No.* 22;—DUNG *laid on* a *clayey* Loam, in *December,*
was of no benefit to *Mixgraſs.* And,

By *Experiment, No.* 23;—COMPOST *laid on* a *clayey* Loam, in *December,* was not beneficial to *Mixgraſs.*

By *Experiment, No.* 24 *to* 28;—COMPOST *laid on* the Surface of a
ſtrong Loam, in *Winter,* was of no ſervice to *Clover.*

By *Experiment, No.* 37;—YARD LIQUOR was of no obvious ſervice
to *Tares* or *Clover.*

By *Experiment, No.* 48, 49;—HERBACEOUS MELIORATION was not
obviouſly beneficial to *Barley.*

By *Obſervation, p.* 17;—COMPOST *laid on* a *ſandy* Loam, or a *gravelly* Loam, in *Winter,* was very beneficial to *Clover.*

By *Obſervation, p.* 40;—COMPOST *laid on* a weak *clayey* Loam, in
January, was *perhaps* of ſervice to *Mixgraſs.*

By *Obſervation, p.* 65;—SUMMER-FALLOWING, alone, gave a good
Crop of *Oats.*

By *Obſervation, p.* 96;—MELIORATION is a moſt abſtruſe department of Agriculture.

* Were I to recommend any preciſe *Form* of Communication, it would be that
which I have adopted for WHEAT 1778.

S

HAVING

MANURE.

HAVING thus taken a fyftematic View of Agricultural VEGETABLES, with the other Elements appertaining to the Vegetable Management *, which occurred in 1777 and 1778; next ought to have fucceeded a fyftematic View of the agricultural ANIMALS during thofe two years. But for Reafons which will be given under the head LIVE-STOCK, I decline the Animal Management wholly in this Publication.

After the *paffive* Elements, or the PATIENTS of Agriculture, fucceed the *active* Elements, or AGENTS of Agriculture. Thefe are divifible into NATURE, or the Agents of *natural* Vegetation; and ART, or the Agents of *factitious* Vegetation; the former of which are fubdivifible into the

Laws of Nature appertaining to Vegetation, or the Vegetable Œconomy.
———————————————— to Animation, or the Animal Œconomy.
———————————————— to the Weather, or Atmofpheric Œconomy.

Nothing however having occurred with refpect to the two former, I will proceed to treat fully of the laft; to which the other two are more or lefs fubject.

THE WEATHER.

HAVING, during the courfe of *two Summers*, recorded a variety of OBSERVATIONS, and regiftered a fhort courfe of EXPERIMENTS appertaining to this ufeful department of human knowledge, I fhould think myfelf blameable were I not to make ufe of this early occafion of communicating them to the PUBLIC. And, in order to be more fully underftood, I will firft give a Defcription of the annexed PLATE appertaining to the WEATHER.

* The Reafon for placing the *Foffile* Elements *after* the Vegetables, here, and *before* them in the Syftematic Index, is the principal Intention of this Publication is to give a fyftematic View of AGRICULTURAL VEGETABLES: Whereas the Index is meant to give a comprehenfive fyftematic View of the Outlines of AGRICULTURE in general.

DESCRIP-

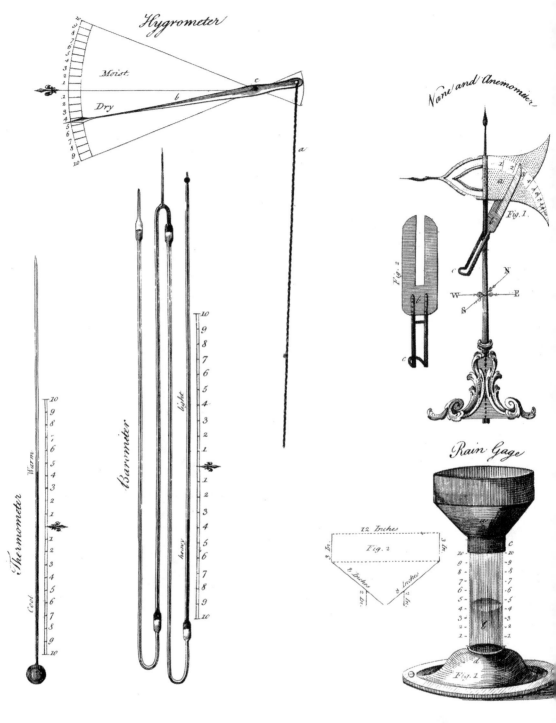

Hygrometer

Moist

Dry

Vane and Anemometer

Fig. 1.

Fig. 2.

N
W   E
S

Barometer

light

heavy

Thermometer

Warm

Cool

Rain Gage

12 Inches

Fig. 2

Fig. 1.

Exhalation Gage

DESCRIPTION of the PLATE.

BEFORE I began to regifter *the ftate of the Atmofphere*, (fee a fpecimen at the clofe of this Article) I confidered how I fhould make myfelf underftood by the Reader who might not be poffeffed of Inftruments exactly fimilar to thofe from which I was about to form my Regifter. After fome reflection, I fell upon an expedient which ftruck me as eligible, and which I flatter myfelf will meet with the approbation of thofe into whofe hands the following ftrictures on the Weather may fall.

I wifhed to hit upon an expedient of being underftood by every man who is poffeffed of a *Barometer* (for inftance), whether it happens to be conftructed with a fingle Tube, and confequently *falls* for rain; or with a double or a quadruple Tube, and *rifes* for rain; or whether it happens to be perpendicular or horizontal, or on the fcale of an Englifhman, a Foreigner, or of any other fcale or conftruction whatever.

The expedient I pitched upon was this: Having, from repeated obfervations during three or four years, afcertained (not intentionally) the *Extent of the Scale* (marked the higheft and the loweft points to which I ever remembered to have feen the Tincture rife or fall), I divided this Extent into two equal parts; confidering the point of Bifection as the point of *Temperature*, or *medial degree of Gravity*, and fubdividing the fpaces, above and below this *Meridian*, into ten degrees each.

Every man who has paid attention to his Barometer, let it be of any of the defcriptions abovementioned, may readily affix to it a *correfponding Scale*; and confequently will be able to fix a fufficiently precife idea to the fubfequent Table, fo far as it refpects the BAROMETER; and by a fimilar rule will, with equal facility, be enabled to underftand the other columns of the Regifter.

It

WEATHER.

It may be faid in objection, perhaps, that few men have paid a fufficient attention to their Barometers to afcertain, with any degree of truth, the medial degree of Gravitude from their own obfervations. If this be juft, let thofe who have not, adopt the degree of *Temperature* which the Maker has marked upon the Inftrument, and let them *fuppofe* the *Extent of the Scale*, until by Experience they have *corrected* and *afcertained* it.

The BAROMETER, THERMOMETER, and HYGROSCOPE, in the Plate, are the Drawings of the Inftruments from which the Regifter was made: The ANEMOMETER, RAIN-GAGE, and EXHALEM'ETER, are theoretic Drawings, from which no Inftruments have as yet been conftructed.

BAROMETER.

A double or quadruple Tube has one advantage over a fingle Tube; the degrees of variation are larger and more *diftinct*; they do not require that *clofe* and minute infpection which a *fimple* mercury Tube does; confequently the Tafk (if it may be called one) of Obfervation is rendered more agreeable. But a double Tube being more fubject to the Influence of the *Heat* of the Atmofphere than a fingle one is, the latter is much more correct, and confequently more eligible. And the annexed Drawing is given merely as being explanatory of the Regifter.

THERMOMETER.

The reafon why the Scale of *Fahrenheit* was not made ufe of was, I thought it more *fimplex* to reduce the feveral Scales of the entire Sett of Inftruments to *one common Principle*. It may be proper to mention, however, that the medial Point of this Scale *happens* to anfwer to the 54th Degree of *Fahrenheit*.

HYGRO-

HYGROMETER.

This conftruction was adopted on account of its *Simplicity*, and, at the fame time, its *Mechanicality* ; (*a*) a hempen Line (called by the Saddlers *Hunting cord*, — a fpecies of large Whip-cord) five feet long ; (*b*) an iron Hand,—Pointer, or Index ; (*c*) the Fulcrum, on which the Index moves. The length of the *Index*, from the *Fulcrum* to the *Point*, is ten inches ; that of the *Lever*, from the *Fulcrum* to the *Eye* to which the cord is fixed, is $2\frac{1}{2}$ inches. Thefe proportions were adapted to the place of its fix-ture ; this from which the Drawing is taken being the firft I made, and was intended merely as an Effay. However, after trying feveral other proportions, I could not hit upon one fuperior to that which chance, as it were, furnifhed me with.

My primary intentions were to have fteeped the cord in *Brine* be-fore I put it up, in order that the *Salt* might more readily have imbibed the moifture ; but on reflecting that this would deftroy the Simplicity I had been looking for ; that as the Salt might *wafte*, the effect would *change* ; and that the abforbency of the *hempen cord* alone was fufficient to anfwer the purpofe intended ; I put it up as it came from the hands of the manufacturer.

The principle on which this Hygrometer acts, appears to be obvious without explanation. The air becoming *moift*, the Cord imbibes its moifture ; the Line, in confequence, fhrinks, and the Index *rifes*. On the contrary, the air becoming *dry*, the Cord dif-charges its moifture, ftretches,—and the Index *falls* *.

It may be true ; that no two Hygrometers will *accurately* keep pace with each other †. I will venture to fay, however, from feven

* See Note (1) to the following Regifter of the State of the Atmofphere.

† There has been one recently invented, which is made from *Hatter's Paper*, and which is faid to excel in this refpect.

months.

months experience, that two Hygrometers on the simple construction abovementioned, have coincided *sufficiently* for the *Uses* of Agriculture. The *Curious*, indeed, might frequently have been dissatisfied, especially when these two Instruments were first put up; but the longer they remained together the more consonant they grew.

It is true, they diminished in the *degree* of action; but as the Scale may be readily diminished in *extent*, and as a fresh Line may be so cheaply and so readily supplied, this is not a valid objection. And indeed this diminution in the degree of action depends considerably on the construction; the propriety, or rather *delicacy*, of which rests almost solely on this point: *The weight of the Index* should be so proportioned to *the weight of the Lever and Cord*, as that the Cord may be kept *straight* without being too much *stretched*. I made one with a *long, heavy Index*; and, in order to gain a more *extensive Scale*, with a *short Lever*: but even when it was first put up it could barely act, and in a few weeks it flagged, and was not able to raise the Index, though the air was uncommonly moist. I therefore made another, with the same *length* both of *Index* and *Lever*, but with a *lighter Index* and a *heavier Lever*, so as to gain the proportion abovementioned; and it has acted exceeding well, and is that which is mentioned to have kept pace with the original one from which the Drawing was made.

Every man who wishes to make use of the HYGROMETER should not have less than *two*. Three or four would be more adviseable; they would then assist in correcting each other; and in case of renewal or alteration, there would be no danger of losing the state of the Atmosphere; which, if only *one* is kept, must necessarily be the Case *.

* The Principle on which this HYGROMETER is formed, is not confined to a small Cord and an Index of ten inches long: it may be extended to a Rope of any length or thickness, and to an Index and Scale of almost any dimensions and extent.

I in-

I intended, and still intend to construct one, or more, on a portable construction. An *Axe* is the form I have pitched upon. The *Edge*, graduated, will constitute the *Scale*; and the *Handle* will receive the *Cord*. This may be hung up in the shade, fully exposed to the action of the air; or, by means of a spike in the end of the Handle, it may be placed in the open Field. *Perhaps* by placing it in a Fallow, it may be actuated by the *Perspiration of the Earth*; *perhaps* among Vegetables, by *vegetable Perspiration*; perhaps by the means of one, or more probably by the means of several, placed at varied heights on the acclivity of a Mountain, the different degrees of moisture at the several altitudes may be ascertained, &c. &c.

I do not mean to speak of the HYGROMETER as any thing *new*; but, perhaps, it has not had that *scientific* attention paid to it which it deserves.

ANEMOMETER.

An instrument to measure the *Velocity*, or rather the *Force* of the *Wind*, has, I believe, been considered as difficult to construct. The shops have them on a variety of constructions: I have not, however, seen any which in THEORY strikes me with *utility*, in a manner equal to that of which I have ventured to give a sketch; although I have not, as yet constructed it.

(*a*) A common VANE (more necessary than an Anemometer to a fore-knowledge of the Weather, and which is essentially necessary to every Farmer). (*b b*) The ANEMOMETER (made of the same materials as the Vane) fixed to the *socket* of the Vane, so as to veer with it freely round the Spindle; and consequently will ever have its *flat* side to the Wind. In a calm, the ANEMOMETER will hang perpendicularly; as the Wind rises, it will of course be forced out of its perpendicular position in proportion to the velocity or power of the Wind, and its degree of force will be pointed out by the graduated Scale on the VANE. The figures being cut out of the

I   Plate,

WEATHER.

Plate, or *Web* of the Vane; or, as in the Drawing, painted black on a white ground; they will be rendered confpicuous. The *Weight* of the *Counterpoife* or *Balance* (*c*), the *Breadth* of the *Web*, and the *length* or *extent* of the *Scale*, muft be in fuch proportion as that, by a violent Gale of Wind, the ANEMOMETER will reach the tenth degree, its outmoft extent*. The only difficulty I can *forefee* is, the ANEMOMETER may in fome degree obftruct the veering of the VANE; but if the VANE be made *large*, and the ANEMOMETER *fmall*, I hope this difficulty will vanifh.

Every part of the Vane, &c. except the acting parts, fhould be made *round*, that it may veer *fmoothly* with the wind.

RAIN-GAGE.

It is fomewhat remarkable, that an Inftrument fo well under-ftood, and which has been frequently ufed, fhould be fo difficult to procure. After going to fome of the principal Inftrument-makers in London, I found a *Copper Tunnel*, the top of which was to be made of fome certain dimenfions, and which I was told was the Rudiment of a *Rain-Gage*; but being informed that the Rain-Water, after it had been collected, was to be *gaged*, or meafured, by fome certain calculation, I concluded to make ufe of a *common Tunnel*; as by a very little *more Calculation* and trouble, the quantity of Rain might be as exactly afcertained as it could be by any other *imperfect* apparatus. I did not, however, reft fatisfied until I had thought of one which might have *fome* claim to perfection. The want, indeed, is fo fimple, and the way to it fo fhort, that it feems almoft difficult to miftake it.

FIG. 1. (*a*) the Top or Tunnel, to be made of copper, iron, or tin, *exactly circular*, and *exactly twelve inches in diameter*. (*b*) A veffel

---

* *Fig.* 1. is a perfpective Reprefentation of the ANEMOMETER fixed to the VANE: *Fig.* 2. a Delineation of the ANEMOMETER alone.

exactly

exactly cylindrical, and *exactly three inches and eight-tenths in diameter* on
the infide. (*c*) A collar fixed to the *Tunnel*, and fitted to the *Cylinder*, fo as to keep the Tunnel *exactly upright*, its top *perfectly level*,
and to prevent its being blown out of the cylinder by the Wind.
(*d*) a foot or bottom, to ftand upon ; which, though broad in proportion to the cylinder, fhould be faftened to a plank, block, &c.
as an apparatus of this kind, being neceffarily placed in an open
expofure, is liable to be thrown down by the wind. *Fig.* 2. is a
Section of the Top or Tunnel.

The Principles on which this Rain-Gage is conftructed are thefe :

The areas of circles being in proportion to each other as the
fquares of their diameters ; and the fquare of the diameter of the
Top, or Tunnel of this Apparatus being 144, the fquare of
the diameter of a cylinder, the area of whofe bafe is *one-tenth*
of the Tunnel, muft confequently be 14.4; the fquare-root of
which is 3.794733, or, for common ufe, 3.8 inches ; confequently,
by merely inferting any ordinary rule or fcale, divided into *inches and
tenths*, into the Rain-Water collected by the Tunnel, and contained
in the Cylinder, the exact *Depth* of rain which has fallen will be
afcertained to the thoufandth part of an inch ; for every *inch* will be
one-tenth, every *tenth*, one-hundredth, &c. *.

An improvement of this Apparatus would be a *glafs* Cylinder;
provided it could be made fufficiently *true*; as by a fixed fcale, the
Quantity of Rain would be known by fimple infpection.

EXHALATION-GAGE.

This is an inftrument which probably has never been conftructed
in any form ; *perhaps*, it has not before been thought of. The intention of it is the reverfe of the RAIN-GAGE ; *this* afcertaining the depth
of *Rain-Water* which *falls* from the atmofphere ; *that* the depth of

* A *black* Scale would perhaps be the moft eligible.

T

naked

naked Water which is *exhaled*, or *abforbed*, by the atmofphere. The foregoing fketch, it is true, is *theoretical*; but it is fo *fimplex* and fo *obvious*, that it can fcarcely fail of being adequate to its intention in practice.

(*a*) May be a *cylindrical*, or any other *parallel* veffel of *any* diameter or *any* depth. (*b*) A glafs Tube, whofe *length* is exactly equal to *ten times* the *depth* of the veffel, iffuing from the *bottom*, and rifing gradually until it reaches a height equal to the *top*, of the veffel.

Perhaps ;—*a cylindrical veffel of twelve inches diameter and* three inches deep, *and confequently furnifhed with a Tube* thirty inches long, *would be eligible dimenfions for an* EXHALATION-GAGE ; by which dimenfions the foregoing Sketch was drawn, on a Scale of *one-tenth* to an *Inch*. Confequently the conftructed Apparatus will be exactly ten times as large as the Drawing.

Thefe dimenfions are preferred for the following reafons : The *Shallownefs* of the veffel fhortens the Tube ; and its *Width* will give the Wind and Sun an opportunity of acting more freely on the Surface of the Water, than a *narrow, deep* veffel would admit of. It muft be obferved, however, that thefe dimenfions are only *theoretical :* EXPERIENCE, alone, can point out the *true* Dimenfions *.

* It may with propriety be afked, " Why are thefe *theoretic Sketches* offered ? Why were not the *Inftruments themfelves conftructed* and *tefted* before they were made public ?" The anfwer is fhort—I have not had an opportunity to conftruct them, nor time to teft them ; but I have both time and opportunity of making them public in this theoretic way, which cannot prevent my attempting to perfect them myfelf ; but may give others an opportunity of doing it in a way I may never think of, and to a degree I may never reach.

Mr. ADAMS, an ingenious Mathematical Inftrument-maker in *Fleet-ftreet* has been kind enough to offer his beft endeavours towards the perfecting of thefe Inftruments, and he will pay a proper attention to fuch hints as may be offered by Gentlemen who may wifh to facilitate his endeavours.

More

More will be faid concerning thefe Inftruments and Apparatus's appertaining to the Weather, in the courfe of the following Obfervations on that fubject.

BEFORE I communicate the Obfervations which I have recently made, I will firft take the liberty of tranfcribing an Extract of a Paper which, in the year 1770, I fent to the *Secretary* of the ROYAL SOCIETY, including an Incident which occurred to me at an early age.

This Paper was written in confequence of an Experiment which was made (or faid to be made) relative to the difparity in the quantity of Rain-Water collected at the top, and the quantity gathered, in the fame time, at the foot, of Weftminfter Abbey: and obferving in the public prints, that this circumftance was accounted for in a variety of ways; fome placing it to the account of *Electricity*, and others, perhaps more *naturally*, to the *waggery of the Weftminfter School-boys*; I did not hefitate to draw up the paper of which the following is an extract.

' It having been found by experience, that the quantity of Rain-water which is gathered at a greater height in the atmofphere is confiderably lefs than that which falls, in the like fpace and time, in a more intimate vicinity with the furface of the earth, I beg leave to lay before the Royal Society an obfervation which fome years ago I made, and which, with fubmiffion to that learned body, I conceive not only to account for THE GENERATION OF DROPS OF RAIN, but alfo ferves to elucidate the Effect abovementioned.

' Thefe Obfervations were made accidentally, about feven years ago, whilft I travelled acrofs a mountainous country on a rainy day; and as I then made a Minute of them, I will beg leave to tranfcribe that Minute, with the Inferences I then drew.

" *Memorandum*

" *Memorandum.* In riding acrofs the Moors between Sinnington and Stokefly, (in the North-riding of Yorkfhire) I made the following Obfervations :

" Before my entrance upon the firft Moor (or Heath), which forms an eafy acclivity, there fell a moderate rain : the drops were rather *large* than *numerous*; and the day, to every appearance, was fet-in for rain *.

" As I approached nearer to the hills, I perceived them to be fheathed in a fog or cloud. While I rode up the eafy afcent, and particularly after I entered the fog, (the Hill which I was then afcending being there pretty fteep) I perceived the Drops to diminifh in fize, and that the higher I mounted the more they diminifhed; until having nearly reached the fummit, I found myfelf furrounded by a fine *equal* Mift, Fog, or Cloud, of floating aqueous Particles, without one perceptible drop of Rain.

" While I continued afcending, and while I rode acrofs the fummit of the Hill, this equal Fog continued; but, on defcending into a Valley on the other fide (this Valley is not deep), I began to perceive fome fmall Drops of Rain : as I defcended, the Drops grew *larger*, and the Mift *rarer*, until having arrived below the Cloud, I found myfelf in the *Rain* afore-mentioned : But, on afcending a fecond Hill, the Drops again diminifhed, until I re-entered gradually the equal Fog.

" The Inferences I drew at the time are thefe : The fmall aqueous Particles which, floating in the air, compofe Clouds, are, by fome *external* Motion, as the WIND (which at this time blew ftrong), or by fome *internal* power, as ATTRACTION, caufed to unite. While they remain in a feparated ftate, the weight of the air is able to buoy them up; but fome two or three of them, let us fuppofe, having be-

' On enquiry, I found that in the Country I had left, it continued to be a rainy day.'

come

come united, and being by that means rendered specifically heavier than the Air, they *begin* to descend But their specific gravity being still *dubiously* greater than that of the Atmosphere, they do not yet descend perpendicularly, but are still driven obliquely by the Wind or other external power; until having fallen upon, and attracted to such other Particles as Chance throws in their way, they descend to the Earth in Drops of Rain.

" This seems to account for the largeness of these Drops during a squall of *Wind*, and after *lightning*; the floating Particles being more strongly agitated under these than milder circumstances. Under the *first*, the descent is rendered more oblique, and perhaps frequently curvilinear, whereby not only the smaller but the larger Drops may become united: Under the *latter*, it is not, perhaps, so much the *external Motion*, as the *Rarefaction of the Air*, which admits of a more easy *precipitation*. The Depth and Density of the Cloud may at other times regulate the size of the Drops of Rain."

' —— It may, perhaps, be objected, that this is not a similar case to that alluded to, and that the top of Westminster-Abbey is seldom hid in the Clouds. To this I answer, that although the intermediate space between the top of the Abbey and the streets of Westminster may not constantly be filled with a *perceptible Fog*, yet I will venture to assert, that it is rarely void of humid particles; and that were the experiment to be tried while that space were filled with a *dry Air* (perhaps this would be most properly done during a summer shower, falling from a cloud floating high in the atmosphere), and again while the same space is filled with a *humid Air*, or Fog, I am fully persuaded that the difference would be essential; and that the Experimentalist would find, from the *former* part of the Experiment, the difference in quantity small, but from the *latter* part very considerable. Indeed to me it seems more than probable,---I will venture to say obvious,---that it sometimes rains in the streets of West-

minster

WEATHER.

minfter, while at the top of the Abbey, inftead of *falling Drops*, the Atmofphere is occupied by *floating Particles* *.

' That the lower part of the Atmofphere generally abounds with *imperceptible Vapour*, and that the Clouds are only a collection of that vapour, feem evident from the quick, or as it were, inftantaneous tranfition from a clear to a cloudy day; efpecially within the Tropics, where I have feen the moft ferene, tranquil day changed in a few minutes to the blackeft tempeft. The Clouds do not feem to rife out of the fea, nor to be brought from a diftant region; but, from an imperceptible beginning, *gather*, until they become livid, and wholly overfpread the vifible Hemifphere; then enfues a violent fquall of Wind with Rain.

' Briefly, the *efficient* caufes of Rain *feem* to be thefe : The particles of matter which form, in their *diffolved ftate* †, the cryftalline fluid WATER, are, as they lie expofed in the fea, or on the furface of the earth, attracted, evaporated, or exhaled into the Atmofphere. Here they float, to us *imperceptible*, until, by fome law of Nature, they are affembled and formed into *vifible* Vapour, or CLOUDS. In this ftate they become more expofed to the Wind, and perhaps other external powers : their degree of vicinity to each may likewife contribute to a change; but, chiefly, THE STATE OF THE ATMOSPHERE, which before kept them afunder, now admits of their union; and thus becoming fpecifically heavier than the Medium which before fupported them, they begin to defcend : in their defcent, they attract to, and fall upon each other, until being formed into Drops of Rain, they return precipitately to their former liquid ftate.'

---

* Is not this an Inference which may be fafely drawn from the Incident above related ? For the depth of the Valley abovementioned cannot, I think, be greater than the height of this building.

† ' While they compofe ICE they are probably in their *natural ftate*.'

Being

Being a total ftranger to the *Orders* of the Royal Society, this Paper was enclofed in a cover addreffed to the *Secretary*; confequently, not having the honour of being introduced by a *learned Fellow*, it could not be *received*. The Incident it contains, however, will, I flatter myfelf, be confidered as interefting to the fubject under confideration, and will fully warrant its being communicated to the PUBLIC in this unlearned way.

I will next beg leave to refer the Reader to the MINUTES on the Weather made prior to July 1777, and which are referred to in page 32 of the DIGEST of the MINUTES of AGRICULTURE, and then proceed to give the OBSERVATIONS made fince that period.

OBSERVATIONS, 1777.

THE following Minutes, among many others, on the WEA-THER, were made during Harveft. They were generally made when CONTRADICTORY PROGNOSTICS prefented themfelves at the fame time; or when fome CHANGE of the Weather feemed to be *po tended*; or, when I was ftruck with fome PROGNOSTIC APPEARANCE which I wifhed to *prove* from my own Obfervation.

AUGUST 6. A *Whirlwind* (*a*); the Sky beautifully mottled with fhell-like Clouds, and with a *deep blue* ground (*b*); — the Clouds *high* (*c*); but the *Barometer* kept getting up (*d*)—and the *Sun* fet foul (*e*).

7. This morning is very fine; the air fwarms with *Infects* (*f*);—the ground is covered with *Cobwebs* (*g*); and *Salt* is as dry as Gun-

(*a*) *Whirlwind* is faid to be a fign of fine Weather.
(*b*) *Dark blue Sky*, the fame.
(*c*) *High Clouds*, the fame.
(*d*) Double Tube which *rifes for Rain*.
(*e*) *Sun fetting foul*, a prognoftic of Rain.
(*f*) *Infects fwarming in the Air*, dubious.
(*g*) *The Ground clad with Cobwebs*, a Sign of fine Weather

4   powder

WEATHER,

powder (*h*). But the *Cock* crows inceffantly (*i*)—and the *Glafs* has rifen an inch and half laft night.

8. *Every thing* portends fine Weather except the *Barometer*, which has kept *creeping up* all day (*k*). The *Sun* fet *dufky*; but the *Clouds* to the eaft are beautifully variegated and *high*.

9. The *Glafs* rofe two inches laft night; the *Wind* got *back* from N. to S. E. and the Atmofphere looked tempeftuous (*l*); but, to-day the Wind has got quite round to the N.; the Glafs has kept falling very faft all day; and there is every appearance of fettled Weather.

This has been a falfe alarm; not only of the *Barometer*, but of the *fetting Sun*, and almoft every other Prognoftic. But perhaps the *Scale of Rain* has been fo much lightened of late (*m*), that the *atmofpheric Balance* now preponderates in favour of fair Weather. And, generally,

Perhaps;—*Settled Rain is prognoftic of fettled fair Weather.*

14. The remarkably fine Weather ftill continues; but, this evening, the *Gnats fing* very much (*n*); and the *Moon* is *foul* (*o*). The *Sun* did not *fet* very well; but the *Canopy* is now (half paft eight) divinely beautiful. It was *loftier* (*p*) this afternoon than I ever remember to have feen it. The *Glafs wavering* (*q*); the *Wind vary-*

(*h*) *When Salt is dry,* a Sign of fine Weather.

(*i*) *The Crowing of the Cock,* in the day-time, is efteemed prognoftic of Rain.

(*k*) *The Barometer creeping up* is efteemed to be a certain Prognoftic of Rain; but I have generally found a *quick rife* to portend it more truly.

(*l*) *The Wind getting back* is efteemed a Sign of foul Weather. It is faid to get *back*, when it moves from N. to W. &c. and *forward*, when it fhifts from N. to E. &c.

(*m*) The Months of June and July were moft remarkably rainy.

(*n*) The *Singing of Gnats* is called a Sign of Rain

(*o*) The *Foulnefs of the Moon* (an annular Cloudinefs round her), the fame.

(*p*) The *Loftinefs of the Canopy* (the apparent concavity formed by the Clouds and the azure Expanfe) is perhaps one of the trueft Prognoftics of fine Weather.

(*q*) The *Wavering of the Barometer* I have frequently obferved to portend Rain.

ing

ing (*r*): the *Geese took wing* (*s*), and flew eagerly into the water, putting it in a ferment.

16. Yesterday it looked threatening; but to day has every appearance of fine weather.

Therefore ;—*The Loftiness of the Canopy* was a truer prognostic than the *singing of Gnats*—the *foulness of the Moon*—the *wavering of the Glass*—*the varying of the Wind*—and the *flying of Geese*, taken jointly.

But perhaps, the Atmosphere is not yet sufficiently replenished with moisture to be actuated by ordinary causes.

28. Great quantities of *Cobwebs on the ground*.

29. A drizzly Morning.—*Geese* flew and cackled much (grateful, no doubt, for the Rain). A very fine afternoon, and a finer evening ;---the *Glass wavering*.

30. The *Glass got up* this afternoon *remarkably fast* (*t*), although there was an exceeding fine *Dew* (*u*); and the *Windows* were very damp on the *inside* (*w*). The *Sun rose orangey* (*x*); the *Clouds sunk* and grew *heavy* (*y*). It began, about half past nine, to *sprinkle* (*z*), and continued till near *eleven* (*a*), when a drizzly Shower set in.

(*r*) The *varying of the Wind* is also said to be prognostic of foul Weather.

(*s*) If *Geese* are *ever* Weather-wise, they have *this year*, unfortunately, been very much out in their Prognostics : and yet, from the repeated Observations I have made, I apprehend they are not this year less wife than heretofore.

(*t*) See Note (*k*).

(*u*) A *strong Dew* is esteemed prognostic of a fine Day.

(*w*) *The Window damp on the inside*---the same.

(*x*) *The Sun rising gaudily*, (the clouds in the east being tinged with an *orange-colour*, called a *gaudy Morning*) is generally esteemed to be a sign of Rain.

(*y*) *The Clouds getting low and heavy*---the same.

(*z*) To *sprinkle* (or *spit*), to rain flow in largish drops.

(*a*) When Rain sets in about *Eleven*, or *between Eleven and Twelve* o'clock, it frequently continues during the afternoon

U   The

The *Wind* got *back* from E. to S. W. — Complaints of *Aches and Pains* (*b*) from all quarters.—It has rained hard all the afternoon.

31. Laſt night and this morning, dry and very windy. Eleven o'clock—*Geeſe* flew and cackled remarkably. A fine Afternoon.

1 SEPTEMBER. A froſty night and a lovely day.

From the whole of theſe Obſervations, it ſeems probable, that *Animals* have no inſtinctive preſcience of the Weather (*c*). They alſo point out clearly, that there are no *certain* Prognoſtics of the Weather; as almoſt every ſuppoſed Prognoſticator has *contradicted* itſelf; excepting that,

A *high Canopy* has uniformly foretold fine Weather;

A *low Canopy* immediately preceded Rain;

A *quick Riſe of the Barometer* portended preſent Rain;

The *Sun riſing gaudily* foretold a rainy Day; and that

Chronic Aches and Pains were prognoſtic of foul Weather.

OBSERVATIONS, 1778.

Monday, 20 JULY. This Summer, *until to-day*, has been uncommonly *dry*.

THE SUN, during this Summer, has almoſt uniformly *ſet well:* ſometimes *clear*, ſometimes among *broken Clouds*, and frequently very *red*. On Friday laſt he ſet *remarkably ſplendid*;—on Saturday, *ſomewhat foul*;—laſt night, *remarkably gloomy:* he ſunk behind a mountain of

(*b*) *Chronic Pains* are often truly prognoſtic. The *Rheumatic Shoulder* of an old Labourer, and the *Corns* of an old Woman, are frequently in the right; yet by no means infallible.

(*c*) But theſe Obſervations are far from being thought deciſive: for it may be ſaid, if *chronical Aches* (which, if influenced by the Weather, become a kind of *accidental Inſtinct*,) be truly prognoſtic in the human ſpecies, why may not ſome other animal have *natural Impulſes* equally preſcient?

livid

livid vapour near ten degrees high, and has not since appeared. This morning, before the Rain set in, he attempted to force his way through the veil which was drawn before him, but without effect.

The Moon is now in *the middle of her last Quarter*; her influence, therefore, had not, probably, any share in this change of the Weather.

The Canopy * has been *uncommonly clear*, until last night, when the *Clouds* encreafed both in number and fize. Last night the Canopy appeared exceedingly *wild* and *tempestuous* (thin, livid *Clouds* flying in every direction, with a pale straw-coloured *ground*). This morning it was almost uniformly *grey*, excepting where the *Sun* attempted to perforate it; and there the *Clouds* appeared *livid*, and the *ground* nearly *white*.

The Barometer has this Summer been uniformly *heavy* †, excepting once, when it fore-shewed a *Thunder-storm*. Yesterday, last night, and this morning, it has got *light* very fast. It now *thunders*, and *rains* very hard.

The Hygrometer on Saturday became *dry*; yesterday, last night, and this morning, *moist*.

The Thermometer has been *hotter* this Summer than perhaps was ever known in this country; and although it now *rains*, the air still continues *warm*.

Wednesday, 29 July. The Weather, after the 20th, was uniformly rainy, until Friday afternoon; which was fine, and was portended by the *setting Sun*. For,

---

* That principally-imaginary Concavity formed by the *Sky* (or rather, *perhaps,* the darkness of Space), the *Heavenly Bodies, Clouds, Meteors,* &c.

† This mode of Expreffion is adopted for the fake of brevity, as well as in conformity to the principles on which the Scales of the foregoing Instruments are formed; and, confequently, it is adopted for the fake of being more *generally* understood.

THE

WEATHER.

THE SUN did not shew himself from Sunday evening until Thurs-day. Thursday afternoon was very gloomy; but just before Sunset the immediate horizon became clear; and the Sun, sinking from behind a very opake cloud, had just time enough to shew the whole of his Disk; setting perfectly *clear* and *splendid*. *Friday* was very *rainy*, until about *three o'clock* in the afternoon, when the Sun broke out exceedingly bright.

HYGROMETER.

HYGROMETERS are *various*; and, when we reflect on the implicit confidence which the Housewife places in her *Salt-box*, the Carter in his *Whit-leather* Thong, and the Sailor in his *Shrouds*, we must acknowledge them to be, in *some* degree, *prognostic of the Weather*.

SIMPLICITY and ACCURACY were the Principles on which I wished to construct an Hygrometer; and, after having thought of many, I pitched upon that described in the Plate.

I put it up about the middle of July, during very hot Weather; and the following are the MINUTES I made respecting it:

Saturday 1st AUGUST. While the *dry* Weather continued, the HYGROMETER remained at about *three* deg. *dry*. During the Rain (see the foregoing Minutes) it got to *three* deg. *moist*, and portended almost every shower. The BAROMETER and the HYGROSCOPE nearly kept pace with each other, until to-day, when the latter is 2½ deg. *moist*, and is still getting *moister*, while the former has sunk considerably, and is still getting *heavier*.

Wednesday, 5th AUGUST. The Weather since Saturday has been very *fine*; yet I observed that some *Pease* which *ought* to have been ready to carry, did not *wither* nor *dry*, nearly in proportion to the *Wind* and *Sun* which they received. Yesterday the Hygrometer fell from *two* deg. *moist* to *below temperate*, when they dried as much in two or three hours as they had before done in a whole day.

The

The HYGROMETER, whether or not it is *prognostic of the Weather*, is a most valuable Oracle to a Farmer. For, to repeat the above-mentioned Incident, yesterday morning while the HYGROMETER stood at *two* deg. *moist*, the Pease of P 3. were by no means fit for carrying; the *Halm* was *green*, and the *Pease* soft. About ten o'clock the HYGROMETER fell to *one* deg. *dry*:—before one o'clock the Pease were in good order. I went up to Adscomb, and took the Teams into the field merely on the *word* of the HYGROMETER, and found the Pease perfectly fit to be carried.

This is a valuable property of the HYGROMETER which I had not foreseen; yet, after being known, it appears sufficiently obvious to have shewn itself in THEORY: for the *Air*, *perhaps*, acts as a *Sponge*; and, while it is *dry*, it *absorbs* the *contiguous moisture*; which, when the Air itself is *moist*, is suffered to remain; or, at least, is not caused to dissipate so rapidly as when the Air is *dry*.

An HYGROMETER, in this point of view, is obvious and simple: for the same Air which absorbs the moisture of the *hempen Cord*, absorbs that of the *Hay* or *Corn*; consequently they will become proportionably *dry* or *moist* according to the degree of *dryness* or *moisture* of the Air. But the *Cord* points out the degree with *mathematical certainty*; whereas, to know whether *Hay* or *Corn* dries *fast* or *slowly*, recourse must be had to *repeated* handling; and, on a straggling Farm, they may be two or three miles distant from the hand.

Therefore;---*on a scattered Farm*, an HYGROMETER must be, *in Hay-time* and *Harvest*, peculiarly useful.

Tuesday, 11th. To-day threatened very much for Rain, and the BAROMETER got *light*; but the HYGROMETER stood at 1⅓ deg. *dry*. It sprinkled in the evening; but the Rain went off, and the BAROMETER fell.

Thursday 13. The BAROMETER has kept getting up all this day. But the CANOPY is *clear*, and the HYGROMETER stands firm at *one* deg. *dry*.   Friday

WEATHER.

Friday 14. During laſt night the BAROMETER has got up *(light)* very much. This morning the SUN roſe *clear*, but was preſently hid by a thick *Scud.*—The HYGROMETER ſtill firm at *one* deg. *dry*. In about half an hour the Scud flew rapidly off, and the CANOPY is now (ſix o'clock in the morning) *perfectly clear*.

The WIND preſently got up, blowing very hard : and I had laid it down as a maxim, that " when the Air gets *light*, but ſtill continues *dry*, it portends WIND ;" and was pleaſed with the apparent diſcovery, that the HYGROMETER was the much-wiſhed for *Check* to the BAROMETER, ſhewing when the latter portends RAIN, and when it foreſhews WIND.

About twelve o'clock the WIND fell, when a drizzly rain ſet in ; which, however, preſently went off. This confirmed me in my opinion ; for I reaſoned thus : " Notwithſtanding the *Lightneſs* of the Air, it does not RAIN, becauſe the Air is *dry*." This reaſoning, however, ſoon became vague ; for, about three (notwithſtanding the BAROMETER, after the WIND ceaſed, fell more rapidly than I ever remember to have obſerved it) a STEADY RAIN ſet in, which ſtill (ſix in the evening) encreaſes ; contrary to *every* Prognoſtic, *natural* and *factitious !*

Saturday 15. THE RAIN ceaſed juſt before Sunſet, which was fine ; and to-day has been a delightful harveſt-day. The BAROMETER and HYGROSCOPE have kept falling all day ; the latter ſtands now (ſeven in the evening) at Temperate.

The Incident of yeſterday is poſitive proof that there are *other cauſes* of RAIN than the *Lightneſs* and *Moiſture* of the AIR : for the ATMOSPHERE was yeſterday neither *light* nor *moiſt* ; yet it RAINED moderately during three hours and a half.

But *perhaps*, when the ATMOSPHERE is *heavy* and *dry*, there is no fear of MUCH RAIN : it is, however, proved beyond contradiction,

that

that a SUMMER SHOWER, or even a TRANSIENT RAIN, does not depend wholly on the *Humidity* and *Gravity* of the ATMOSPHERE.

Yesterday's Shower came at right angle to the Wind. The Wind was S. W. and the Rain came from the N. W. or N N. W.

Tuesday 18. This forenoon the HYGROMETER sunk to *five* deg. *dry*. The Oats which were mown yesterday; and which, in patches, were weedy, are; to-day, as dry as tinder; the succulent Weeds though as *green* as when cut, may be rubbed to powder.

Wednesday 19. This morning was *foggy*. During the FOG, the HYGROMETER got to *three* deg. *moist*; but, the Fog passing off, it is now (four in the evening) near *ten* deg. *dry!* It is observable, that the SUN is remarkably *scorching*; although the THERMOMETER is only *five* deg. *warm*.

Perhaps, the Air being *dry*, the Rays of the Sun, meeting with no obstruction, fall *unblunted* on the skin. In the *Shade* it is barely *pleasant*.

———————————

By the foregoing INCIDENTS and REFLECTIONS I was led on to a *closer Investigation* of the WEATHER.

The first step I took was to make out a

PROGNOSTICAL ARRANGEMENT.

After having made out a miscellaneous List, gathered from my own Observations, from popular Maxims, from *Dr. Campbell*'s ' Shepherd of Banbury,' &c. &c. I *classed* them in the following order.   **I**

WEATHER.

PROGNOSTICS OF SEASONS*.

Spring, rainy; ferene Summer.
January, mild; bad for Grafs Crops.
February, fair; wet Seed-time, bad Crops.
Feb.-Mar. rainy; wet Spring and Summer.
Feb.-Mar. dry; dry Spring and Summer.
April, windy; good Crops.
April, fhowery; good Crops.
May, cold and windy; good Crops.
May, wet; bad Crops.
Summer, hot and dry till the middle of Sept. cold Spring.
Summer, moift and cool; hard Winter.
Summer, ferene; windy Autumn.
Autumn, ferene; windy Winter.
September, hot and dry; cold Spring.
Oct.-Nov. warm and rainy; Jan. and Feb. frofty and cold.
Oct.-Nov. fnow and froft; January and February mild.
Winter, warm and open; hot and dry Summer.
Winter, windy; rainy Spring.
Winter, commencing with a fouth Wind; cold Winter.

* The reafon for making this diftinction muft be obvious to every one who has made obfervations on the Weather. Every feafon has its peculiar *Characteriftic*; it it either *wet, dry,* or *changeable*; and from the above Minutes, as well as from repeated, unminuted Obfervations, I am clearly of opinion, that the fame Prognoftics, which in a *fair Seafon* will bring *fair,* will, in a *rainy Seafon,* bring *Rain.* Therefore the Characteriftic of the *Seafon* ought to be pointed out before the Quality of the *Weather* can be prognofticated.

4

Winter,

WEATHER.

Winter,   commencing with a north Wind ; mild Winter.

Wind,   easterly the fore-part of Summer ; dry Summer.

Wind,   westerly the latter part of Summer ; dry Autumn.

Wind,   a week's fair Weather with a southerly Wind ; Drought.

Lightning,   without Thunder ; Drought.

Cuckoos,   coming early ; hot Summer.

Woodcocks, ————— ; cold Winter.

Fieldfares, ————— ; cold Winter.

Acorns,   plentiful ; hard Winter.

Haws,   ——— ; hard Winter.

Nuts,   ——— ; good Harvest.

Broom,   full of Blossoms ; plenty.

Almond,   full of Bloom ; plentiful Harvest.

Walls,   sweating much in the beginning of Winter ; open Winter.

PROGNOSTICS OF THE WEATHER.
SUN.

Rising orangey ; Rain.

Rising red and fiery ; Wind and Rain.

Rising cloudy, if the Clouds decrease ; fair.

Rising dim ; drizzly.

Rising with his eyes half open ; Rain.

Setting foul ; Rain.

Setting red ; Wind or Rain.

Setting blue ; Rain.

Setting dusky, and streaked with red ; Storm.

Setting purple ; Fine.

Setting bright ; Fine.

Setting behind a dusky Bank ; Rain.

X

MOON.

WEATHER.

MOON.

Bright, with sharp Tips; Fair.

Dim, with a Ring round her; Rain.

New Moon not appearing till the 4th day; rainy Month.

The lower Horn of the New Moon sullied; foul Weather before the Full.

The middle ———; Storms about the Full.

The upper Horn ———; foul about the Wane.

On the 4th Day, bright, sharp, and oblique; fair Month.

——— ——— dim, blunt, and erect; Rain.

Saturday's Moon; rainy Month.

ATMOSPHERE.

Cold, after Rain; Rain.

Cold, in Summer; Rain.

Warm, in Winter; Rain.

Sultry, in Summer; Thunder.

Heavy; Fair.

Light; Rain.

Moist; Rain.

Dry; Fair.

CANOPY.

Red in the Evening and grey in the Morning; Fair.

Green between the Clouds; Stormy.

High; Fair.

Low; Rainy.

Orangey in the Morning; Rain.

Deep blue ground; Fair.

Pale blue ground; Rainy.

Lined

Lined with Streamers; Rain.
Speckled; Fair.
Refembling the inverted Cavity of a Ship; fettled Fair.

CLOUDS.

Small and white; Fair.
Streaming within the Canopy; Rain.
Large like rocks or towers; Showers.
If fmall clouds increafe; Rain.
If large ones decreafe; Fair.
Large Clouds with white tops and livid bafes; Thunder
Two fuch Clouds rifing on either hand; fudden Tempeft.
Small, ragged, livid Clouds near the Sun, efpecially when
    fetting; Rain.
If Clouds increafe rapidly; Tempeft.
Fleecy Clouds fcattered from the eaft; Rain within three
    days.
Waterifh Clouds on the tops of Mountains; Rain.
Mountains free from Clouds; Fine.
High and light; Fair.
Low and heavy; Rain.

RAINBOW.

In the Morning; Rainy.
In the Evening; Fine.
The Fruftum of a Rainbow; Rain.
Predominantly red; Wind.
——————— green or blue; Rain.

WIND.

Whirlwind; fettled Fair.
Continuing in the N. E. three days without Rain; Fair for
    eight or nine days.

Going

WEATHER,

Going backward ; Rain.

Unfettled ; Rain.

From N. W. to N. E. Fair.

From S. E. to S. W. Rainy.

RAIN.

Sudden Rain feldom lafts long.

Coming on by degrees, is likely to laft fix hours.

Beginning with a high foutherly Wind, and the Wind falls ; Rain for twelve hours.

Beginning before Sun-rife will end before Noon.

A fhower before Sun-rife ; a fine day.

Beginning an hour or two after Sun-rife ; a rainy day.

Cold Wind after Rain ; more Rain.

Setting in between eleven and twelve o'clock ; a rainy afternoon.

Clearing up about that time ; a fine afternoon.

A fquall of Rain or Hail ; fettled Fair.

A rainy Friday ; a rainy Sunday.

A rainy Sunday ; a daggly week.

MIST or FOG.

A mifty Morning, and the Mift fall ; a hot day.

———————— and the Mift rife ; Rain.

If general before Sun-rife near the Full ; fine Weather.

———————————— in the new Moon ; Rain in the old.

———————————————— old Moon ; Rain in the new.

DEW.

A heavy Dew ; Fair.

None ; Rainy.

And if it vanifh fuddenly, or early ; Rain.

FROST.

A white Froft ; Rain within three days.

SEA.

WEATHER.

SEA.

Suddenly fwelling; Tempeft.
Making a hollow Noife; Storm.

RIVERS.

Sinking more than ufual; Rain.

MINES.

Certain Fumes rifing; a Storm.
The Water becoming muddy; a Storm.
An Eruption of certain Meteors; a Storm.

SOUNDS.

Heard diftinctly from a great diftance; Rain.

SMELLS.

Unufual and difagreeable; Rain.

FEELINGS.

A difagreeable Languor; Thunder.
The Hand feeling dry; Rain.
Chronic Pains being more violent than ufual; Rain in
Summer, or Froft in Winter.

ANIMALS.

Sheep,	feeding during Rain; a continuance of Rain.
Geefe,	wafhing themfelves more than ufual; Rain.
———	cackling and flying much; Rain.
Ducks,	wafhing and diving much; Rain.
Cock,	crowing much in the middle of the day; Rain.
Peacock,	fqualling more than ufual; Rain.
Kites,	flying and hovering high; Fine Weather.

Owls,

WEATHER.

Owls,   screaming in the evening ; Fair and Frost.
Larks,   rising high and singing long ; Fine.
Redbreasts singing loud in the open air ; Fine.
————— faintly under cover ; Rain.
Swallows, skimming the surface of water ; Rain.
Dogs,   the Nose dry ; Rain.
Moles,   earthing more than usual ; Rain.
Frogs,   appearing of a golden hue ; Fine.
————— ———————— dark, dusky colour ; Rainy.
Spiders,   crawling abroad ; Rain.
————— webs on the ground, or in the air ; Fine.
Ants,   moving their Eggs to dry places ; Rain.
Worms,   creeping in numbers out of the ground ; Rain.
Glow worms, appearing more than usual ; Fine.
Bees,   staying much about home ; Rain.
————— flying far abroad, and staying out late ; Fine.
Flies,   peculiarly troublesome ; Rain.
Gnats,   playing in the open air in the Evening ; Heat.
————— ——————— shade ; Showers.
————— stinging much ; Cold and Rain
Fleas,   biting unusually ; Rain.

VEGETABLES.

Burnet,   shutting its Flowers ; Rain.
Dandelion, expanding its Down ; Fair.
Trefoil,   the Stalks swelling and becoming erect ; Rain.

FOSSILS.

Salt,   dry ; Fair.
————— damp ; Rain.
Walls,   damp ; Rain.

This

This Arrangement, however, by no means anfwered my wifh : inftead of ftrengthening my judgment, it led me into a labyrinth apparently endlefs. While I *collected* it, I was difgufted with Maxims
which I knew to be falfe ; and with others which appear vague at firft
fight. Nothing could have induced me to have *claffed* it, much lefs
to have *publifhed* it, had it not been for the purpofe of recording a
System of Popular Maxims, which have more or lefs engaged the
attention of Mankind for upwards of two thoufand years.

It is not my prefent intention to enter into an analytic Inveftigation
of the Weather ; yet I cannot refrain from communicating fuch Obfervations and Reflections as have occurred to me, and which, I flatter
myfelf, will be confidered as at leaft an *Overture* towards founding
a Foreknowledge of the Weather on the durable bafis of
Science. It would, indeed, be deferving of a harfher term than Wantonnefs, for any man even to *attempt* to throw down a work of ages,
without erecting, or at leaft laying the foundation of another in its
ftead, which he is *convinced* will fill its place with greater propriety.

Left this, however, fhould be thought to be leaving the fubject too
abruptly, I will fubmit the following Theory in corroboration of thofe
principles which I have drawn from Experience.

Occult Qualities being exploded, it will readily be admitted
that the Weather is influenced by Physical Causes. Can Animals,
Vegetables, and Fossils, be the Causes of Rain ? Can they fend
forth the aqueous Particles ? affemble them in the Atmofphere, or call
them down in Rain ? No man will venture to anfwer in the affirmative.
Why then confider them as Prognofticks of the Weather ? It may
be anfwered, " They are *actuated* by the Caufes." This will, for a
moment, be admitted. But *fuppofing* them to be *actuated*, why apply
to them ? Why fhall we not inveftigate the Caufes themfelves, and
look up to thefe for information ? It will *not*, however, be impli-

*

citly

citly granted that they *are actuated* with any degree of *Uniformity* and *Certainty*. For to what end could this Law of Nature be inftituted? Can it be traced to the felf-prefervation of the individual, or to the propagation of the fpecies? *Suppofing* fuch an *Inftinct* to exift in the ANIMAL Creation, it is moft likely to be the ftrongeft in the defencelefs tribe of *Infects*; and yet I have frequently feen Swallows feeding high in the Atmofphere, not only immediately before, but during the beginning of Rain, which has afterwards lafted for fome time. This is prefumptive evidence, at leaft, that even Infects have not a prefcience of Rain. After they have been *actuated* by the *Effect of the Rain*, felf-prefervation, no doubt, leads them to the earth for fhelter.

It may, however, be faid, that " VEGETABLES are not, like ANIMALS, fubject to the influences of *Fear*, *Hunger*, and *Luft*; their impulfe muft of courfe proceed fimply from the Caufes of Rain." *If* there are Vegetables which, with *Regularity* and *Certainty*, fhut up their organs of fructification on the *Approach* of Rain, fuch Vegetables are truly prognoftic of *immediate Rain*. But, admitting this, fo fhort a notice as they hold out can only be of momentary utility, while the leading Caufes of Rain may long precede the effect. Befides, Flowers are *tranfient* even to a proverb, and while exiftent muft be *fought for* in the field: Whereas, *factitious Prognoftics* are ever ready for infpection. And is it not confiftent with Infinite Wifdom, that thefe feeble impulfes which the Brute and Vegetable Creation *may* be fubject to, were intended to point out to Mankind that the *ftate of the Atmofphere* is *liable to changes?* and confequently, to excite them to the exertion of thofe facultie, with which they are fuperiorly bleffed, in the *invefigation of thofe changes?*

Having thus briefly apologized for venturing to deviate from the eftablifhed principles of Prognoftications, I will endeavour to explain

plain more fully thofe on which I wifh to found a foreknowledge of the Weather.

THE WEATHER, taken in its largeft fenfe, is THE STATE OF THE ATMOSPHERE : confequently, TO PROGNOSTICATE is TO FORESHEW THE STATE OF THE ATMOSPHERE.

THE STATE OF THE ATMOSPHERE is influenced or modified by its variations of Denfity, or WEIGHT ; by the quantity of aqueous Particles it contains, HUMIDITY ; by the Invifibility of its contained Particles, termed FINENESS ; by the vifible Collections of thofe Particles, CLOUDINESS ; by a precipitation of thofe Particles, RAIN, &c. by its MOTION, or *Wind* ; by FROST ; by its different degrees of HEAT ; by LIGHTNING, &c.

To fimplify the enquiry as much as poffible, and to endeavour to render it of *ordinary Ufe*, I will confine it folely to RAIN ; neverthelefs admitting fuch other *Accidents* of the Atmofphere as refult from this Enquiry.

In the Atmofphere, as in the Animal Syftem, a CIRCULATION is kept up ; the *evaporable Particles* which lie expofed on the furface of the Globe are abforbed by the contiguous air, in which they float as *invifible Vapour*, or foar as *vifible Clouds* ; out of which they return as *Rain*, to the Earth ; from whence they may be again exhaled, again formed into Clouds, and again precipitated in Rain.

The firft enquiry, therefore, towards a PHILOSOPHICAL TREATISE OF THE WEATHER is,---In what manner is the Atmofphere fupplied with HUMIDITY ? The next,--What *caufes* and what *prevents* this INVISIBLE HUMIDITY fufpended in the Atmofphere from being formed into VISIBLE CLOUDS ? And the laft,--What *prevents**, and what *caufes* the VISIBLE CLOUDS to be precipitated in RAIN ?

* For we frequently fee them hover over our heads for days without *falling*.

Thefe

WEATHER.

Thefe, however, are difquifitions I do not mean to enter into; they are myfteries of Nature which perhaps never have been, perhaps never may be *fully* difclofed to human perception. However, whilft the memories of NEWTON and LINNÆUS are recollected, Nature may be purfued in his moft intricate Mazes, with at leaft fome *hope* of fuccefs. And, by an acutenefs of obfervation, aided by a courfe of Phyfical Experiments, the PHILOSOPHY OF THE WEATHER may happily be rendered as obvious as the NEWTONIAN PRINCIPLES, or the LINNÆAN SYSTEM.

Waving therefore the PRIMARY CAUSES, we will proceed to the more obvious MEANS, or EFFICIENT CAUSES, of Rain; and endeavour, by *afcertaining* the PRESENT, to *prognofticate* the APPROACHING, STATE OF THE ATMOSPHERE.

Firft, It will fcarcely be difallowed that the Atmofphere, in its moft dry and tranfparent ftate, contains a portion of humid Particles, and that *vifible* Vapour or Clouds are formed from thefe *invifible Particles*†: confequently, Clouds cannot be formed without pre-exiftent Moifture in the Atmofphere. And may we not fay, that the greater number of aqueous Particles there are in the Air, the greater probability there is of Clouds being formed, and of Rain falling? This being admitted, *the Degree of Moifture in the Atmofphere* feems to be the firft ftep towards a foreknowledge of RAIN.

The ftate of the Atmofphere, with refpect to its MOISTURE, is pointed out by the HYGROMETER: and had we a Hygrofcope which could fhew with *mathematical Certainty* the Degree of Moifture at *every Altitude*, we fhould be forward on our way towards CERTAIN PROGNOSTICATION. But FACTITIOUS HYGROMETERS are placed too near the Earth to point out with certainty the degree of Moifture contained in that region of the Atmofphere in which the *Clouds* are formed: for by an Incident mentioned in page 20, the degrees of

† See page 126.

Humidity

Humidity were various, according to the degrees of height in the Atmofphere *.

The all-bountiful Author of Nature, however, has furnifhed us with NATURAL HYGROMETERS, which, perhaps, if properly attended to, may be nearly adequate to our wants;---I mean, the RISING AND SETTING LUMINARIES. And even the Luminaries in a ftate of afcendency may be confidered as HYGROMETERS. Perhaps, a bright, *fcorching* SUN;--the Moon lucid, with fharp horns;--and the STARS twinkling, denote a DRY AIR. On the contrary, the SUN and STARS faint;--and the MOON blunt; as alfo the folar and lunar RINGS, may indicate a MOIST AIR.

Secondly, It is evident from long-continued Experience, that the *Denfity* or WEIGHT of the Air influences, in fome degree, the

ATMOSPHERIC

* It has, however, been already hinted, in the *Defcription of the Plate*, that by placing FACTITIOUS HYGROMETERS at different heights on the acclivity of a mountain, the Degrees of Moifture might be afcertained at different Altitudes: But as *cultivated* countries are feldom mountainous, and as a mountain may attract an *extraordinary* degree of Moifture, I will here offer other HINT, which may not only be of more general Ufe, but which may alfo afford more decifive Information. As, firft, to place one HYGROMETER at the foot, another at the top, of a very high tower, or other building;—or, which would be ftill lefs fubject to the influence of *attraction*, fufpend an HYGROMETER to an artificial KITE or EAGLE, and raife it high in the Atmofphere;—or, to reach a ftill greater height, fend up an ATMOSPHERICAL HYGROMETER from the top of a tower or high building;---or, to compafs a yet greater fpace, and to gain a ftill greater variety, place one Hygrometer at the foot of a hill; a fecond at the foot of a tower raifed on the fummit of the hill; a third at the fummit of the tower; and fix a fourth to an EAGLE to be raifed from thence. The ROYAL OBSERVATORY, at Greenwich, would afford this variety.

Perhaps, an engine or apparatus of this kind, by whatever name it may be called, whether a *volātus*, an *Aerius*, or an EAGLE, may be confiderably improved; and, if judicioufly made,---by an ingenious workman,---of a proper fize,--from fit materials,---and with fuitable *Machinery*, it may perhaps be raifed to almoft any given

Y 2   height

WEATHER.

ATMOSPHERIC MOISTURE; yet it is at the same time evident, from Incidents which occur in the foregoing MINUTES, that the Gravitude of the Atmosphere is not the *only* efficient cause of RAIN. These Incidents, however, prove that the weight of the Air has a *principal Share* in the government of the Atmosphere.

The State of the Atmosphere, with respect to its specific GRAVITY, is ascertained, with mathematical nicety, by that admirable factitious Prognostic, the BAROMETER.

Third. From the foregoing Minutes, and from common observation, it is evident that the MOTION of the Atmosphere is an efficient Cause of RAIN *. The motion of the Atmosphere is divisible into its DIRECTION, and its VELOCITY, or *Force.*

The State of the Atmosphere, with respect to its VELOCITY, is ascertained by the ANEMOMETER; with respect to its DIRECTION, by the VANE.

---

height. And, *perhaps,* not only the Degrees of MOISTURE, but those of WEIGHT, HEAT, and even MOTION of the middle regions of the Atmosphere, might, by such a machine, be in some measure ascertained.

It may be said in objection, that an instrument of this nature could not be of *common use*; because it could only be raised during a strong Wind. In reply it may be said, that after CURIOSITY has been *satisfied,* and the ATMOSPHERICAL APPEARANCES have been ascertained, the repetition will no longer be required.

* When we consider the MOTION of the Atmosphere in its largest sense, every other *prognostic state* seems to be borne away with it; for that which is the local Atmosphere of any given place to-day, may to-morrow be removed to a considerable distance; and, if the Wind remain stationary, may in a few days become the local Atmosphere of a far distant country. However, while the causes and the *laws* of Wind (if the term may be applied to a subject so apparently *lawless*) remain little understood, we ought not, perhaps, to rest too implicit a confidence on that Motion which may happen to take place in the immediate Region of our Observation. Nevertheless the MOTION OF THE ATMOSPHERE, *near the Earth,* may, I think, be safely considered as one of the principal efficient causes of Rain.

1

Fourthly.

Fourthly. Perhaps the HEAT of the Atmofphere (whether it be communicated by the *Sun* or the *Earth* *) may be confidered as an efficient Caufe of RAIN.

The State of the Atmofphere, with refpect to its HEAT or *Warmth*, is afcertained (near the Earth at leaft) with a great degree of Precifion by the THERMOMETER.

Laftly. Perhaps the QUANTITY OF RAIN *recently fallen*, regulates in fome meafure the QUANTITY OF RAIN *prefently to fall*.

The ftate of the Atmofphere, with refpect to its *Exhauftion* by the QUANTITY OF RAIN recently fallen, is afcertained, with mathematical certainty, by the RAIN GAGE.

Thus we have enumerated five *Influencers of the ftate of the Atmofphere* ; namely, its

HUMIDITY,	HEAT,
GRAVITUDE,	and
MOTION,	RAIN.

And we have likewife enumerated fix FACTITIOUS PROGNOSTICS, whereby *the degree of Influence* may be afcertained ; namely,

The HYGROMETER,	ANEMOMETER,
BAROMETER,	VANE,
THERMOMETER,	RAIN-GAGE.

---

* VULCANOES are proofs of the exiftence of SUBTERRANEOUS FIRE. And what are ufually termed GROUND THAWS (thaws which do not feem to proceed from any change in the Atmofphere) are prefumptive evidence that fubterraneous Heat is fometimes communicated to the Atmofphere thro' the pores of the Earth. Might not this communication be tefted by the affiftance of Thermometers, placed at different depths in Caverns or Mines? Is it not probable that the upward *Progrefs* of the Heat (for its paffage cannot be fuppofed to be *inftantaneous*) might be traced, by comparing thefe Thermometers placed at different depths ; and its emiffion be confirmed by fuch as are placed on the furface of the Earth?

The

WEATHER.

The five latter may be termed *perfect* Inftruments : the firft is very *imperfect* ; as the degree of moifture may vary at different altitudes. But in order to regulate this defect, I have pointed out, in the courfe of thefe obfervations, a NATURAL HYGROMETER ; namely, the APPEARANCE OF THE ATMOSPHERE, or rather of the ATMOSPHERIC CANOPY ; whether that appearance be formed and altered by the

CLOUDS,	LIGHTNING,
SUN,	RAINBOW,
MOON,	or the
STARS,	AURORA BOREALIS.

Neverthelefs, the afcertainment of the quantity of aqueous, of *rainy* Particles fufpended in the Air, remains the moft indefinite part of this enquiry ; and yet, perhaps, to afcertain this with fome degree of precifion is effentially neceffary to PROGNOSTICATION.

There might perhaps be an Apparatus conftructed which would affift, confiderably, the natural and factitious Hygrometers above-mentioned, in afcertaining the *prefent ftate of the Atmofphere*, with refpect to its MOISTURE. Were it not for the MOTION of the At-mofphere, it might, with the affiftance of the RAIN-GAGE, bring this part of the fubject to a confiderable degree of certainty* ; I mean, *an Inftrument to afcertain the Quantity of aqueous Particles ex-haled by the Atmofphere from naked Water.*

Laft Summer I fketched out fuch an Inftrument, and have given an Engraving of it in the preceding Plate, under the name of an EXHALATION-GAGE.

Whether

---

* And when we confider that the fame caufe which fends forth Exhalations in any certain place, may extend its influence to a confiderable diftance ; and when we re-

fiect

Whether or not this Apparatus may prove inftrumental to Prognostication, it will go near to demonftrate the cause of Exhalation: whether it is the *drynefs of the* Air;—the *weight of the* Air;
the *Heat* or *Attraction of the* Sun;—the *Direction* or *Velocity of the*
Wind;---*Subterranean Heat, &c.* *.

IT was the middle of Auguft laft when I digefted my ideas on
this fubject; and having determined to throw afide natural
Prognostics (excepting thofe abovementioned), I began, on Tuefday the 19th of Auguft, an *experimental Regifter of the State of the
Atmofphere,* on the principles here fet forth; namely, by regiftering
(at *Sun-rife, Noon,* and *Sun fet*) the Weight, Moisture, Heat,
Motion, and Appearance of the Atmofphere.

flect that *changeable Winds* return as it were the fame Atmofphere, the Motion of
the Atmofphere may not be fo inimical to the afcertaining of its Moisture, as
may at firft fight appear.

* Befides, it might prove of fome importance, to know the proportion which the
Inhaustion of the Atmofphere by *Exhalation,* bears to its Exhaustion by
Rain; and thereby to afcertain, in a courfe of years, the quantity of *Animal, Ve-
getable,* and *Foffile Perfpiration,* &c. &c.

I cannot quit this part of the fubject without mentioning another advantage which
we may expect from an Exhalation-Gage. I find among the rough Minutes of
the Weather of 1778 the following memorandum: " Perhaps there is an *imperceptible
communication* between the Atmofphere and the Earth: Perhaps the aqueous particles not only rife imperceptibly, but fall again without being formed into Clouds
or Rain. This is evident in *Dews.* And if there is not an invifible communication,
in what fuperior Region lies the *Atmofpheric Ocean?* Where is that vaft abundance of
watery Particles which muft have been exhaled during this almoft rainlefs Summer?
for the *Air,* by the Hygrofcopes natural and factitious, is ftill *dry.* If this conjecture
be well-founded, how vague is the Subject of Prognoftication!" Whether it be well
or ill founded, will, I flatter myfelf, be fully proved by the Exhalation-Gage

STATE

WEATHER.

STATE of the ATMOSPHERE,

From 19 August, to 15 September, 1778.

N. B. h. is an Abbreviation of *heavy*; l. *light*; m. *moist*; d. *dry*; w. *warm*; c. *cool*;
E. *east*; N. *north*, &c. The *Figures* point out the *degree* of Warmneſs, Coolneſs,
Moiſture, Motion, &c. o o. *temperate*; o N. o S. &c. *calm*; the Wind having
gone down at N. or S. the *Vane* ſtanding in that direction. See the Plate and its
Deſcription.

August, 1778.		Weight.	Moisture.	Heat.	Motion.	Appearance.	Quantity of Rain.
19. Wed.	Sunriſe	6. h.	1. m.	4. w.	2. E.	a thick Fog — — —	⎰
	Noon	6. h.	2. d.	5. w.	2. SE.	perfectly ſerene or clear — —	⎬ 0. 00.
	Sunſet	5. h.	6. d.	5. w.	0. E.	cloudleſs, but ruddy (*) — —	⎱
20. Thur.	Sunriſe	5. h.	3. d.	3. w.	0. E.	foggy — — —	⎰
	Noon	5. h.	1. m.	4. w.	1. E.	cloudleſs (1) — — —	⎬ 0. 00.
	Sunſet	5. h.	4. d.	5. w.	0. NE.	ruddy, and the Sun large (a) —	⎱
21. Frid.	Sunriſe	6. h.	3. d.	3. w.	0. NE.	a ground Fog ; but the Canopy clear	⎰
	Noon	6. h.	4. d.	5. w.	3. E.	cloudleſs — — —	⎬ 0. 00.
	Sunſet	5. h.	4. d.	5. w.	2. SE.	Sun ruddy, Canopy high, but watery	⎱
22. Sat.	Sunriſe	5. h.	2. d.	4. w.	0. SE.	foggy (a New Moon at Eight o'clock)	⎰
	Noon	5. h.	0. o.	5. w.	1. E.	a few Clouds — — —	⎬ 0. 00.
	Sunſet	5. h.	1. d.	5. w.	1. E.	Canopy clear ; the Horizon foul —	⎱
23. Sun.	Sunriſe	6. h.	0. o.	4. w.	2. E.	cloudleſs — — —	⎰
	Noon	6. h.	0. o.	5. w.	1. NW.	tranſparent, grizzly Clouds — —	⎬ 0. 00.
	Sunſet	7. h.	1. d.	5. w.	2. NE.	perfectly clear (New Moon not viſible)	⎱
24. Mon.	Sunriſe	8. h.	2. d.	4. w.	1. N.	cloudleſs — — —	⎰
	Noon	8. h.	0. o.	5. w.	3. N.	cloudleſs — — —	⎬ 0. 00.
	Sunſet	8. h.	1. d.	5. w.	2. NW.	a gaudy Canopy (Moon viſible and ſharp)	⎱
25. Tueſ.	Sunriſe	9. h.	1. d.	4. w.	0. SW.	a ground Fog, with a clear Canopy	⎰
	Noon	8. h.	0. o.	6. w.	3. N.	cloudleſs — — —	⎬ 0. 00.
	Sunſet	8. h.	2. d.	7. w.	3. N.	Can. clear, Hor. hazy, Sun ruddy (Moon blunt)	⎱
26. Wed.	Sunriſe	8. h.	2. d.	3. w.	2. N.	ruddy before ; red at — —	⎰
	Noon	8. h.	1. d.	4. w.	3. NE.	a clouded grizzle (b) — —	⎬ 0. 00.
	Sunſet	7. h.	2. d.	4. w.	7. NE.	Canopy clear ; Horizon foul and watery	⎱

(*) *Ruddy* ; a purpliſh bloom, which ſometimes appears near the ſetting Sun.

(1) During the Fog, the Hygrometer roſe from *ſix* Deg. *dry* to *two* Deg. *moiſt*. When the Fog went off, it got down again to *four* Deg. *dry*.

(a) The *Largeneſs* of the riſing or ſetting Sun probably denotes a *moiſt* Air.

(b) The Sun was preſently hid with a N. E. Blight, which broke into large Clouds about Eight. Cold and threatening all day.

August, 1778.		Weight.	Moisture.	Heat.	Motion.	Appearance.	Quantity of Rain.
27. Thur.	Sunrise	8. h.	3. d.	2. w.	o. NE.	Canopy dusky; Horizon clear —	
	Noon	8. h.	4. d.	2. w.	6. NE.	cloudy and threatening — —	0. 00.
	Sunset	8. h.(c)	3. d.	2. w.	2. NE.	Canopy cloudy; Horizon foul (d).—	
28. Frid.	Sunrise	8. h.	1. d.	2. w.	2. NE.	cloudless — — —	
	Noon	8. h.	4. d.	3. w.	3. NE.	large Clouds — — —	0. 00.
	Sunset	9. h.	3. d.	3. w.	2. NE.	Canopy cloudy; Horizon clear —	
29. Sat.	Sunrise	9½!	3. d.	1. w.	o. NE.	a transparent Fog — — —	
	Noon	9. h.	3. d.	2. w.	2. NE.	cloudless — — — —	0. 00.
	Sunset	8. h.	2. d.	3. w.	o. NE.	clear and serene — — —	
30. Sund.	Sunrise	6 l.	2. d.	2. w.	o. N.	cloudless and very bright — —	
	Noon	5. l.	2. d.	3. w.	4. NW.	large Clouds — — —	0. 05.
	Sunset	4. l.	2. d.	2. w.	6. NW.	tempestuous (e) — — —	
31. Mon.	Sunrise	6. l.	3. d.	1. w.	o. NW.	bright and cloudless — —	
	Noon	7. l.	3. d.	2. w.	5. NW.	some clouds — — —	0. 00.
	Sunset	6. l.	3. d.	1. w.	o. NW.	red, with a few Clouds — —	
1. Tuef.	Sunrise	6. h.	4. d.	o. o.	o. NE.	clear (The first white Frost) —	
	Noon	5. h.	3. d.	1. w.	3. N.	large clouds — — —	0. 00.
	Sunset	5. h.	3. d.	1. w.	o. N.	Canopy wild; Hor. gloomy and ruddy	
2. Wed.	Sunrise	5. h.	2. d.	1. w.	1. NW.	a hazy morning. About 7 a drizzle set in, which	
	Noon	4. h.	2. m.	2. w.	3. N.	went off about 9. Before 12 began to drizzle again: soon went off. Afternoon threatening.	0. 00.
	Sunset	4. h.	1. m.	2. w.	1. NW.	Sun set red and gloomy (f) — — —	
3. Thur.	Sunrise	4. h.	1. m.	1. w.	2. NW.	the morning grey. The Sun rose obscure.	
	Noon	4. h.	2. m.	3. w.	o. NW.	Broke out about 9. Fine day. About	0. 07.
	Sunset	4. h.	3. m.	3. w.	o. NW.	6 began to drizzle, and lasted till 8.	
4. Frid.	Sunrise	5. h.	1. d.	1. w.	1. NE.	a fine clear morning — —	
	Noon	4. h.	1. d.	2. w.	2. N.	an exceeding fine day — —	0. 00.
	Sunset	5. h.	2. d.	3. w.	1. NE.	the Sun went down obscure — —	
5. Sat.	Sunrise	5. h.	3. d.	1. w.	2. NE.	the Sun rose clear, with a very opake	
	Noon	5. h.	3. d.	2. w.	4. N.	Canopy, which presently broke into	0. 00.
	Sunset	5. h.	4. d.	2. w.	o. N.	fleecy Clouds. A fine day. Sun set clear.	

(c) The Wind blew *Six* or *Seven* degrees all the middle of the day, without any indicative alteration of the BAROMETER.

(d) By the *foulness* of the Horizon is meant that *Bank* of dusky Vapour frequently seen at sun-set, resembling a far distant mountain, and which probably indicates a *moist* air.

(e) *Tempestuous*; large and livid Clouds, with slender watery streams underneath them.

It is observable that this morning was very fine,—the forenoon cloudy,—the afternoon threatening, with a shower about sunset; after which the Wind returned, and the air again became *heavy*. Altho' the BAROMETER kept getting *one* degree *light* every six hours, the HYGROMETER never varied. The BAROMETER, therefore, without the HYGROSCOPE, only indicated *Wind with a shower*; but not a fall of *Rain*.

(f) During the first two weeks of this Register, I took the *appearance* which presented itself merely at sunrise, noon, and sunset; except when something observable occurred, and then I minuted it as an extra observation. But during the last two weeks, instead of taking the *appearance* as it *happened* just at *Noon*, I have given a sketch of the changes of the appearance of the Atmosphere during the day; which, tho' not so *methodical*, is, I think, more perspicuous, and sets the information in a more interesting light.

Z

WEATHER.

SEPTEMBER, 1778.		Weight.	Moisture.	Heat.	Moisture	Appearance.	Quantity of Rain.
6. Sund.	Sunrise	5. h.	2. d.	0. 0.	2. N.	a fine grey morning — — —	000.
	Noon	4. h.	4. d.	2. w.	4. NE.	the afternoon bleak — — —	
	Sunset	4. h.	2. d.	3. w.	6. NE.	the evening fine and clear — —	
7. Mon.	Sunrise	4. h.	0. 0.	1. w.	4. NE.	a dull bleak morning, followed by very large snow-like Clouds, with a dark blue ground or sky; accompanied with flying showers. The Sun set wild.	0. 05.
	Noon	3. h.	1. m.	1. w.	6. NE.		
	Sunset	3. h.	0. 0.	2. w.	5. NE.		
8. Tuef.	Sunrise	5. h.	1. m.	2. w.	2. NE.	a fine clear morning. A pleasant mild day, with some slight flying showers. The Sun set broken (g)	0. 04.
	Noon	5. h.	1. m.	2. w.	4. NE.		
	Sunfe	5. h.	2. d.	2. w.	2. NE.		
9. Wed.	Sunrise	6. h.	1. d.	2. w.	0. NE.	gloomy, foggy morning. About 10 the Fog rose and broke into large Clouds, with a dark blue Sky. A fine mild day. The Sun set obscure (h)	0. 00.
	Noon	6. h.	1. d.	3. w.	3. NE.		
	Sunset	5. h.	2. d.	3. w.	0. NE.		
10. Thur.	Sunrise	4. h.	1. d.	3. w.	0. NE.	a fine grey morning. The day mild as Spring. The Sun went down broken and foul (i) — — —	0. 00.
	Noon	3. h.	1. d.	3. w.	3. NE.		
	Sunset	2. h.	0. 0.	2. w.	1. NE.		
11. Frid.	Sunrise	1. h.	0. 0.	2. w.	3. SE.	the morning very dull. About 10, began to rain very hard, before a few violent claps of thunder. Rained till half past 1. A fair afternoon. Sun set red and cloudy (k) — —	0. 49.
	Noon	0 0.	0. 0.	3. w.	0. S.		
	Sunset	0 0.	0. 0.	2. w.	1. S.		
12. Sat.	Sunrise	0 0.	0. 0.	2. w.	3. SW.	a fine clear morning. The day threatening; continued fair, here: but some flying showers about. A fine evening. The Sun went down clear. —	0. 00.
	Noon	0 0.	0. 0.	3. w.	5. SW.		
	Sunset	0 0.	1. d.	2. w.	0. SW.		
13. Sund.	Sunrise	3. h.	2. d.	2. w.	1. NE.	a fine bright morning. The middle of the day mild and gloomy; but on the whole a pleasant day. The Sun set broken and red.	0. 00.
	Noon	3. h.	2. d.	2. w.	2. NE.		
	Sunset	4. h.	1. d.	1. w.	0. NE.		
14. Mon.	Sunrise	5. h.	2. d.	0. 0	0. N.	a thick Fog, which fell about 9 —	0. 00.
	Noon	5. h.	1. d.	2. w.	1. NE.	a remarkably serene, pleasant day —	
	Sunset	6 h.	2. d.	2. w.	2. NE.	the Sun set clear and ruddy — —	
15. Tuef.	Sunrise	7. h.	2. d.	0. 0.	0. NE.	a thick Fog, which fell again about 9 —	0. 00.
	Noon	6. h.	0. 0.	2. w.	2. NE.	a lovely fine day — — —	
	Sunset	6. h.	0. 0.	2. w.	1. NE.	and a delightful evening. — —	

(g) *Broken*; partly obscured by small, *broken*, ragged Clouds.

(h) *Obscure*; entirely hid by large *common* Clouds.

(i) *Broken and foul*; partly hid by *broken* Clouds; and the parts which were visible rendered *dim* by the foulnefs of the Atmosphere.

(k) *Red and cloudy*; large Clouds tinged with red.

THE

THE reasons for not continuing this Register farther than the middle of September, are these: The more we simplify an enquiry, especially one so abstruse as is this under consideration, the sooner, probably, we shall arrive at some degree of truth. The Operations of Agriculture which are more particularly subject to the Weather, are principally confined to three Months in the Year; namely, from the middle of June to the middle of September. A foreknowledge of the Weather in other parts of the Year is of far less estimation. Would it not then be embarrassing the enquiry, if it were extended indiscriminately to every part of the Year? May not the same state of the Atmosphere which in Winter prognosticates fair weather, in Summer be succeeded by Rain? To mention one instance: In Winter we seldom have much Rain while the air continues *cold:* In Summer we may have showers, but we rarely have a fall of Rain while the air remains *hot*. This is not mentioned as an invariable rule; but is given in evidence that there may be distinct Prognostics adapted to the different parts of the Year. Why then should we risk this deception? Besides, there is scarcely any man, let him be ever so assiduous, who is equal to the task of circumspectly registering the state of the Atmosphere, Morning, Noon, and Evening, for three hundred and sixty-five Days successively. Whereas, on the other hand, there is scarcely any man who is interested in the operations of Nature, and engaged in the processes of Agriculture, to whom such a Register, during the three months above-mentioned, would not be an agreeable amusement. In Hay-time and Harvest, every Cloud, every change of the Wind, and alteration of the Barometer, is an interesting incident which fixes the attention. And who will deny, that a Man who has an amusement in the execution of a Plan is more likely to succeed, than one to whom the performance is labour? When we have reduced a foreknowledge of the Weather

in

in Hay-time and Harvest to some degree of Certainty, we may then, indeed, presume to work by a larger scale*.

A formidable Objection to this *confined* mode of enquiry naturally presents itself; for it has been already said, and with a great share of truth, that the same Prognostics which in a *rainy Season* produces *rain*, will in a *fair Season* produce *fair*; how therefore are the SEASONS to be *prognosticated* by confining the Register to th ee months in the year? I am sorry to be under the necessity of saying in reply to this Objection, that the PROGNOSTICATION OF SEASONS to any degree of certainy, is, I am afraid, placed wholly out of human reach. When we consider how difficult it is to-day to prognosticate the WEATHER of to-morrow, what hope have we of foretelling in Winter what SEASONS we shall have in the ensuing Spring and Summer? Until we are better informed with respect to the WEATHER, it seems presumptuous to think of prognosticating the SEASONS; especially when we consider the *continued series* of *dry Seasons* which have happened since July 1777, and which still continue. That we have not in this Country any *regularity of Seasons* is evident from the Spring-like Winter and the Summer-like Spring of this year. And I will venture to say that, at present, we have no PROGNOSTICS OF SEASONS which merit the smallest degree of attention from Reason and Common-sense.

Another objection is immediately started : If different Seasons

---

* Although these strictures on the Weather are more particularly applied to AGRICULTURE, yet it is hoped that the basis on which they are raised, is ample enough to admit of their being extended to every other USEFUL SCIENCE. Many advantages no doubt would accrue from an accurate Register, *during the Year*; nevertheless, I think it would, at present, be dangerous, to the *Farmer* at least, to draw, indiscriminately, *general Conclusions*.

have

have their peculiar Prognoftics of the Weather, and if this difference in Seafons cannot be forefhewn, how are we to come at any
degree of certainty with refpect to a foreknowledge of the Weather?
I hope it may fafely be replied, *Prognoftics* of Seasons are in a great
meafure *unneceffary* to a foreknowledge of the Weather, provided
we can afcertain, with any degree of precifion, the *Diagnoftics* of
Seasons; I mean, the difcriminating characteriftics or *Symptoms*
which *attend* a dry, a wet, or a changeable Season: and this
perhaps is all that can be of real ufe towards a Prognostication of
the Weather.

It would no doubt be advantageous to the Hufbandman to have
a foreknowledge of Seasons; efpecially with refpect to the *time of
Sowing* and the *quantity of Seed* (fee Observations, page 55); as
alfo to regulate his *number of pafturing Stock, &c.* And I have
arranged in the foregoing lift fuch Maxims as I was able to collect
on this fubject; though I will candidly confefs myfelf too great a
Sceptic in popular Prognoftics to place any dependance upon them*.

The only Prognostics of Seasons which have ever occurred to
me as in any degree *rational,* are the Quantity of Rain, and the
Motion of the Atmosphere. It feems reafonable that the Quantity of Rain which fell in the paft Seafon, fhould regulate, in
fome meafure, the quantity to fall in the enfuing Seafon. However,
the *continued Series* of dry *Seafons* which have happened within the
laft two years, are prefumptive proof that this is not a *certain* Prognostic of Seasons; for altho' an uncommon quantity of Rain fell
in the Spring and former part of the Summer of 1777; yet that
quantity, perhaps, was by no means adequate to the dry Weather
we have fince had. The Quantity of Rain may neverthelefs
influence, in fome degree, the Characteristics of Seasons; and

---

* I do not mean by this to intimate that there are no Prognostics of Crops:
I fpeak of the Seasons as they are characterized by the State of the Atmosphere.

as the attention required in afcertaining it, efpecially· by any one who is poffeffed of a mathematical RAIN-GAGE, is inconfiderable, it certainly merits that attention, *throughout the Year* *.

With refpeċt to the MOTION OF THE ATMOSPHERE, or WIND ; the *fudden Changes* may not only influence the WEATHER, but its *ſtationary tendency* may, in fome meafure, regulate the CHARACTER OF THE SEASONS ; and notwithſtanding a *minute Regiſter* of its *momentary variations thro' the year* would be a laborious taſk, yet the obfervance of its *principal changes* would require only a fmall ſhare of attention.

Briefly,—the following appear to be the objeċts which the AGRICULTOR ought, at prefent, to hold in view, with refpeċt to a Prognoſtication of the Weather ; namely, to afcertain more precifely the DIAGNOSTICS OF SEASONS † : To keep a MINUTE REGISTER of the ſtate of the Atmofphere, *during the Harveſt-months* : and, To regiſter the QUANTITY OF RAIN, and the STATIONARY WINDS, *during the Year.*

Having thus taken a general view of a Regiſter of the State of the Atmofphere, and as I flatter myfelf the foregoing fpecimen may prove an inducement to others to keep one on the fame, or

---

* Any Gentleman who has been fo fortunate as to have regiſtered the Depth of Rain which fell in the former part of the Summer of 1777, would be doing an effential fervice to this part of the enquiry to make it public ; efpecially if the Regiſter has been continued down to this time (April 1779). The quantity was, no doubt, uncommonly great ; and might, perhaps, bear fome proportion to the dry Weather which has fince happened, and confequently was prognoſtic of it.

† Perhaps, the BAROMETER remaining ſtationarily *heavy* ;—the natural HYGROMETERS indicating a *dry* Air ;—the CANOPY *high* ;—and the WIND *north-eaſterly* ; are DIAGNOSTICS of SETTLED FAIR. On the contrary, the Atmofphere remaining, for feveral days, *light* and *moiſt* ; the CANOPY *low*, with the WIND fixed at *fouth-weſt*, are perhaps, in this country, DIAGNOSTICS of a WET SEASON.

on

on fimilar principles, I will proceed to explain more fully its feveral olumns.

WEIGHT. The *fractional parts of Degrees* are purpofely omitted, for the fake of perfpicuity. " One Degree heavy," or " three Degrees light," conveys immediately a clear and precife idea ; but " one Degree and three quarters heavy," or " one Degree and forty-five Minutes heavy," or " one Degree, feven tenths, and five hundredths heavy, &c." is deftructive of *Perfpicuity*, without adding an Iota to *ufeful Accuracy*.

MOISTURE. This Column muft not be confidered as giving the Moifture of the Atmofphere, *taken generally*, but its degrees of Moifture, *near the Earth*: however, as there may be a *uniform fimilarity* or a *uniform difparity* between the lower and the middle regions of the Air, this may prove a ufeful Column in a Regifter of the State of the Atmofphere. By comparing it with the column APPEARANCE, this *parity* or *difparity* may, perhaps, be fortunately difcovered.

HEAT. The PLATE, and what has been faid under the head WEIGHT, fufficiently explain this Column. I cannot, however, refrain from remarking here the exact coincidence of the *Meridian*, or medial degree of Heat, which I laid down *folely* from my own obfervation (fee page 115.), and the point of *Temperature* which is fubjoined to the Scales of *Reaumur* and *Fahrenheit* *.

MOTION. The *direction* was taken from a well-veering VANE, placed in a good expofure. The *fractional Points* are here omitted, as the *fractional parts of Degrees* are in the other Columns, and for the fame reafon ; for, to a man unexpert in *boxing the Compafs*, N. N. E. by N. (for inftance) would require fome reflection before

* Since writing the above, I have been informed, that the point of Temperature marked on thefe Scales was afcertained by placing a Thermometer in a Cave, where the variation of Heat was fmall; and the exact point was fixed upon by taking the medial degree of Heat in the Cave, in the fame manner as I took it in the open Air.

3   a com-

WEATHER,

a competent idea could be affixed to it; whereas N. E. conveys, immediately, an adequate idea to every man. I have therefore taken the *nearest* direction to the eight principal points of the Compass; with a full conviction that this is all which can, *at present*, be *useful*.

Being destitute of an ANEMOMETER, the *velocity* was *estimated* by its effect on the senses, by the motion of the Trees, &c. The *Scale of Motion*, from a calm to a hurricane, was supposed to be divided into ten degrees : Thus 0 signifies *Calm*; 1 and 2, *a gentle Breeze*; 3 and 4, *a moderate Breeze*; 5 and 6, *a brisk Wind*; 7 and 8, *a strong Wind*; 9 and 10, *a violent Gale*.

This Column, therefore, so far as it respects the *velocity* of the Wind, must not be considered as *accurate*; and an ACCURATE ANEMOMETER, applicable to *common use*, would be a valuable acquisition to PROGNOSTICATION.

APPEARANCE. This Column is exceedingly difficult to register; the appearances of the Atmosphere are innumerable, and the terms whereby to describe those appearances exceedingly confined. The RISING AND SETTING SUN alone give a variety of distinct appearances, each of which ought to have a separate *technical* name assigned it, so that the precise idea might be registered and communicated. The same may be said of the appearance of the CLOUDS; and any man who would analyse the ATMOSPHERICAL APPEARANCES, and assign to each of the most striking a distinct and expressive term, so as an adequate idea of each might be readily and clearly communicated, would be rendering an essential service to this interesting subject.

Perhaps, the most *scientific* manner of doing this would be to catch the different appearances with the *Pencil*, and affix to each DRAWING a concise term or name, expressive of the Appearance it represents. This, to a *philosophic* mind, would be at once a rational and a delightful employment.

QUAN-

QUANTITY OF RAIN. Not being able to procure a mathematical RAIN GAGE, and not having time to wait for the conftruction of one, I was under the neceffity of making ufe of a lefs
fimple, tho' fufficiently accurate, apparatus. It confifted of a common Tin Tunnel to *collect* the Rain-Water; a common glafs bottle
to *receive* it; and a fmall China veffel to *meafure* it in. And, as Gentlemen who refide at a diftance from the Metropolis may find it difficult to procure a mathematical apparatus, yet may wifh to afcertain
the *depth of Rain*; and the method I hit upon being fimple, and may
not immediately occur to every one, I will here fully defcribe it.

Firft, I found the *Area* of the top of the *Tunnel*, thus: Its *mean
Diameter* I found to be 6.7 Inches; the fquare of which is 44.89.
And the *Square of the Diameter* of a circle being in proportion to
its *Area*, as 1 is to .7854. I ftated the following proportion: As
1 is to .7854, fo is 44.89, the *fquare of the Diameter* of the Tunnel, to 35.256606; which is confequently the *Area* of the TUNNEL.

Next, I proceeded to find the *Area* of the Bafe, or end of the
CYLINDER. Its *mean Diameter* I found to be 2.85 inches; the
fquare of which is 8.1225. I then faid, As 1 is to .7854, fo is
8.1225 to 6.3794115, the *Area* of the CYLINDER.

Having once afcertained this *proportion* between the two veffels,
the *Depth of Rain* was any time readily found; for the quantity of
Water collected being placed in the Cylinder, and its depth therein
afcertained, it only remained to fay, As 6.38 * is to that depth, fo
is 35.25 (in the *inverfe*, or *reciprocal* proportion) to the real depth
of Rain which had fallen.

I will exemplify this in the quantity which fell on the 11 September. Having poured the Water, collected in the bottle, into
the Cylinder, I found, by inferting a *Diagonal fcale of equal parts*

* Ufing 6.38 inftead of 6.3794115, fhortens the Operation, without affecting its
accuracy.

A a

(a com-

WEATHER.

(a common box-fcale divided into *inches and tenths*), that the depth was exactly 2.7 inches. I therefore faid (in the inverfe proportion) As 6.38 is to 2.7. fo is 35.25 to .49 ; the depth of rain which fell on the 11th of September.

However, it is hoped, as the RAIN-GAGE defcribed in the Plate is fo fimple as to be conftructed with fufficient accuracy by almoft any ingenious Mechanic, that no man will *rifk* an *error in calculation,* which every man is liable to.

Having thus ventured to explain the foregoing fpecimen as a MODEL, I will now proceed to fpeak of it as a REGISTER.

It muft not be expected that any thing *conclufive* can be drawn from *one Month's* Obfervation ; and more efpecially from the obfervation of a Month *fo uniformly fair* as was that in which the above Regifter was formed; there being only *one variation* during the whole month. However, as it is from Incidents like that of the *11th of September* we muft hope to profit by a REGISTER OF THE STATE OF THE ATMOSPHERE, let us take a view of the feveral Columns prior and fubfequent to that Day.

This half inch of Rain was obvioufly prognofticated by the WEIGHT of the Atmofphere, eight-and-forty hours before it fell; the Air becoming regularly *lighter*, during that time, at the rate of about one degree every eight hours. Neither the MOISTURE, nor the HEAT of the Atmofphere gave any ufeful indication ; nor did its MOTION vary until the morning of the Rain. But the APPEARANCE of the SETTING SUN portended it for three Days ; more efpecially the evening before the Rain. The fhower was attended with fome violent claps of THUNDER *.

To clofe the incident here, a conclufion might be drawn very much in favour of the BAROMETER and the SETTING SUN; but if

* The Thunder was not *heard* until immediately *after* the Rain fet in. The *Lightning,* however, and *diftant* Thunder, might neverthelefs have *preceded* the Rain.

I   we

we continue it to the next day, we are embarraffed; for the *fame
ftate* of the Atmofphere (fome difference in the appearance of the
Setting Sun excepted), which on Friday brought half an inch of
Rain, on Saturday gave a fair day! Nor does any rational conjecture offer itfelf, why a continuance of the fame ftate of the Atmofphere fhould not have brought a continuance of Rain; except
we fuppofe that the Moisture of the Atmofphere received an *extraordinary Stimulus* by the Thunder (or rather the Lightning*)
which attended the Rain; and the Air being by this means *fuddenly
exhaufted* † of its *principal* Moisture, the Rain of courfe ceafed.
We may extend this conjecture, by faying, that the Setting Sun
on Friday denoted fome *remaining* Moisture; which Moifture being
exhaufted by the fhowers which on Saturday fell in different parts
of the country; and, on Sunday, the Atmofphere becoming heavy,
and the Wind getting round to N. E. the Weather, of courfe, returned to settled fair ‡. Thefe,

* Perhaps, had not the Atmofphere received this *fhock*, it might have retained
its Moisture ftill longer, and its Weight and Motion changing fuddenly after,
this Rain might not have fallen. Admitting this conjecture, a foreknowledge
of Lightning is effentially neceffary to a foreknowledge of Rain. And if
Lightning and Electricity are the fame, ought not the Electricity of
the Atmofphere to make a Column in the Regifter of the Weather? It is true,
that a terrestrial Electrometer would be liable to the fame *circumfcription* as
the terrestrial Hygrometer; but might not an Atmospheric Electrometer be raifed in the manner hinted for the atmospherical Hygrometer?
and might it not be productive of afcertaining electrical Appearances, as well
as the latter may be of afcertaining humid Appearances?

It may be *prudent* (I hope it is not *neceffary*) to obferve, that the *Reflections* contained in this and feveral other *Notes* on the Weather are *merely theoretical*; but I
flatter myfelf they will be confidered as *Reflections* which *refult immediately* from *actual
Obfervation.* And although in themfelves they convey nothing *conclufive*, they may
neverthelefs ferve as *Subjects* for future inveftigations; and they are purpofely thrown
into Notes, that they may not interfere with the prefent outline.

† The Rain was uncommonly heavy.

‡ It is obfervable, that the *rifing of the Fog*, on Wednefday, feems to have affifted

very

Thefe, however, are only *probable Inferences*, and no POSITIVE CONCLUSION muft be hoped for from a *fingle* Incident : yet if we are enabled to reafon with fome degree of confiftency on *one*, what may not be expected from a *variety* of *fimilar* Incidents ? and if *one Month* affords us an interefting Incident, what may not be expected from *twenty Years* accurate obfervation ? I do not mean the obfervation of *one* man, in *one* place, but the joint efforts of many in a variety of places. What beneath CERTAINTY can be the refult ? I mean, that *twenty years* accurate Obfervations, made in different parts of the Ifland, will either afford fome CERTAIN PROGNOSTICS, whereby future generations will be confiderably benefited; or it will be known with *Certainty* that a FOREKNOWLEDGE OF THE WEATHER is a myftery of Nature, placed wholly beyond human reach. And I hope I may be permitted to fay, without incurring any unfavourable imputation, that the bare *attempt* will reflect honour on the country it may be made in.

Having thus fubmitted the SELF-EXPERIENCE I have hitherto had with refpect to the Weather ;—having given a PROGNOSTICAL ARRANGEMENT of *popular Maxims* relative to this fubject ;—and having ventured to make an overture towards SCIENTIFIC PROGNOS-TICATION, I will conclude this article with offering up my moft earneft wifhes, that Gentlemen of leifure and obfervation may bend their attention towards this ufeful and interefting fubject ; and that Men of Science may unite their efforts in refcuing this important department of human knowledge from the hands of *vulgar Error*, and in founding a Foreknowledge of the Weather on the PRINCIPLES OF SCIENCE, and the unvarying LAWS OF NATURE *.

very much in giving *foulnefs* to the Atmofphere; whereas, the *falling of the Fog*, on the Monday and Tuefday following, rendered it uncommonly *pure*.

* For abftracts of the Weather of 1777 and 1778, and the effects which the dif-ferent Seafons had on agricultural Vegetables, fee the foregoing OBSERVATIONS, page 17, 22, 31, 35, 46, 76, 90, 96.

SERVANTS.

SERVANTS.

NOTHING interesting with respect to this branch of Management having occurred to me since the *publication* of the MINUTES OF AGRICULTURE; and as I have there communicated my sentiments fully on this subject, I will beg leave to refer the Reader to page 35 of the DIGEST: and can only add here, that I have not yet found any reason to deviate from the outlines of management I have there attempted to draw.

BEASTS OF LABOUR.

NEARLY the same may be repeated of this article. I will therefore here only observe, that I am still clearly of opinion, that OXEN are equal to *every* department of Agriculture; and that whatever FARMING HORSES *do*, OXEN, of a proper size, and properly managed, *can do.* Yet I am still of opinion, that the working of OXEN will not in this country become *general*, until some *restraint on the breeding of horses* shall take place, or the WORKING OF OXEN be promoted, by some extraordinary exertion of PUBLIC SPIRIT, or PUBLIC MUNIFICENCE; and that it is an object of NATIONAL IMPORT need not here be urged.

IMPLEMENTS

IMPLEMENTS.

WHETHER my Invention was *wearied*, or whether it was *satisfied*, with refpect to AGRICULTURAL IMPLEMENTS, I am not a Judge: I can only fay, that fince finifhing the BURYSOD SWING-PLOW, I have not found myfelf defirous of encreafing the number of Inftruments of Hufbandry.

Time and *conftant ufing* are the only tefts of farming utenfils ; and I can with the ftri&teft truth declare, that after one, two, three, or four years *common ufe*, I have found WOODEN OX-COLLARS,—BURYSOD SWING-PLOWS,—CONCAVE HINGE-HARROWS, — DOUBLE HAND-HOES, and LAND-PLANES, fuperior to any other Implements *I have ever made ufe of* for the purpofes for which they are feverally intended. And without any other motives than that honeft fame which every man has a right to afpire at, added to an earneft defire of promoting, to the utmoft of my abilities, the Agriculture of this country, I will not hefitate to *recommend* them to the attention of every *Farmer*, to whofe foil and fituation they are adapted: namely, the BURYSOD SWING-PLOW, I will venture to recommend for every foil which requires to be *acclivated*, or raifed into *Ridges* :—the DOUBLE HAND-HOES, for *Sandy Loams* :—the LAND-PLANE, for *light Soils*, and for every Soil or Crop which requires a *fmooth Surface*, more efpecially for *Leying* .--and the WOODEN OX-COLLARS, for every Soil and Situation in the Ifland ; a fharp *flinty* Soil *perhaps* excepted.

Engravings and Defcriptions of thefe Implements are given in the DIGEST of the MINUTES OF AGRICULTURE : And they may be procured by applying to Mr. SHARP, *Leadenhall-ftreet*, or at his Manufactory in *Tooley-ftreet, Southwark*.

DIVISION

DIVISION OF FARMS.

HAVING fpoken fully of this Article in the MINUTES OF AGRICULTURE, and as no frefh materials, except an Experiment on PLASHING HEDGES, have occurred fince July 1777, I muft beg leave to refer to the DIGEST of the Minutes, page 68.

I cannot, however, refrain from faying here, that I have fufficiently experienced the conveniency of CLASSING *fmall arable fields* (fee the PLAN OF THE FARM) and confidering each DIVISION as one *entire Field*, to *recommend* it to others. Indeed its utility is fo felf-evident as fcarcely to admit of recommendation.

By Exp. No. 29; *Trimming the Plafhers*, and filling in with *naked Rods*, gave a better Fence than plafhing the Quick *rough*, and filling in with *fprayey Boughs*.

And a LIVE RODDLE HEDGE, by the fide of a DITCH adapted to the Soil it is made in, is, perhaps, the *Ultimate* of Farm hedge-making; for it is not only an immediate Fence againft *every* fpecies of pafturing ftock, but, if kept *trimmed*, will remain fuch during the ufual terms of a leafe.

To fee ditches overflowing with every fhower, and hedges look ruffetty in June, is difgraceful to a civilized country; efpecially to one which ftands fo forward in the Improvements of Agriculture as does this country. But, until the Farms of *arable* Fields are rendered lefs *crooked*; the Ditches of *wet* land Farms are made *deeper*; and the Hedges throughout the kingdom wear a verdure in Summer, let us not *boaft of Superiority*, either in the ufes or the ornaments of Agriculture.

How convenient it would be to the Plowman,—how advantageous to the Farmer,—and how ornamental to the face of the country, were Farms univerfally divided into *fquare Fields*, with *neatly trimmed live Hedges*. By *trimming* is not meant that the Hedge fhould necef-

farily

farily be rendered *low* : On *grazing* Farms, where *Shelter* is wanted, they fhould only be trimmed *on each fide* ; fo as to render the upper parts denfe *(thick* it cannot properly be called), and to prevent the bottom fhoots from being deftroyed by the dripping of the upper fpray; whereby the bottom of the Hedge foon becomes naked and open, and is confequently no longer a Fence againft the fmaller agricultural animals.

Thofe who plant *Trees in hedges*, by way of *fhelter*, deceive themfelves, and punifh their pafturing ftock; for inftead of *breaking off* the wind, the trees *add ftrength* to it (as the Piers of a bridge encreafe the rapidity of a river); and the brute which flies to them for protection, is fubjected to additional feverity. A Quick Hedge trimmed in fuch a manner as to be rendered *thick at the bottom*, is a fhelter for cattle nearly equal to a wall of fimilar height.

SUCCESSION.

THE Succession of Crops (or rather of the Occupants of the Soil, whether Crops, or Fallow) may be *regular* or [*irregular*.

The aboriginal Farmers in general obferve *no regular Rotation*; but crop their fields with what judgment or caprice points out. A *regular Succeffion*, however, is not a modern invention; as many of the Common-field Lands in England have been fubjected to it from time immemorial.

That a REGULAR ROUND is *convenient* cannot be denied; and that it ought to be adopted by every *young Farmer*, is equally evident : however, a man of mature judgment *may perhaps* tread a more *profitable* path by a lefs mechanical guide.

Turnips,

Turnips, Barley, Clover, Wheat, is at prefent the *Fafhion.* Some Land-lords, it is faid, have been fo *high* in it as to infift upon inferting a claufe in their Leafes that their Tenants fhould inviolably obferve this Rotation; without confidering whether their eftates would produce Barley, or bear the Fold! This, however, was being fomewhat too *lordly*; for altho' a *regular* Succeffion may be preferable to a *confufed* one, yet it ought to be adapted to Soils and Situations. And, perhaps, the only general Rule which can be given refpecting it, is to adopt fuch a ROTATION as will keep the Soil *clean* and in *Heart,* and at the fame time afford a fupply of fuch Vegetables as are beft adapted to the *Soil, Climature, Sale,* and *Confumption,* taken jointly.

By Experiment, No. 8 ;---SUMMER-FALLOW, WHEAT—gave a larger Crop and a cleaner Quondal than HORSE-HOED BEANS, WHEAT—did.

By Experiment, No. 11 ;---WHEAT fucceeding FALLOW, BEATEN ROAD,—VIRGIN EARTH,—BEANS, or TARE-BARLEY, was a good Crop.

By Obfervation, p. 18 ;—FALLOW, CLOVER—was good management.

By Obfervation, p. 32 ;---TARES, TARE-BARLEY—was ineligible.

By Obfervation, p. 41, 42 ;---FALLOW, MIXGRASS—was good management.

Obfervation, p. 60;---A comparative View of FALLOW, WHEAT, and CLOVER, WHEAT.

By Obfervation, p. 66 ;—SUMMER-FALLOW, OATS, CLOVER—was, on a *clayey Loam,* moft hufbandly management.

By Obfervation, p. 67 ;---WHEAT, OATS, CLOVER—was, on the fame Soil, very bad management.

See Obfervations, page 68, 69 ;---on the *ftiff-land* SUCCESSION.

By Obfervation, p. 77 ;---SUMMER-FALLOWING for CLOVER was eligible management.

By Obfervation, p. 82 ;---SUMMER-FALLOWING for LEY GRASSES was eligible.

B b

By

SUCCESSION.

By Obfervation, p. 86 ;---*Perhaps,* to produce a *hard, benty Hay,* frequently *mow* ; to gain a *foft, herby Hay,* frequently *pafture.*

By Obfervation, p. 97 ;-----FOUL CLOVER - LEYS are unfit for WHEAT.

By Obfervation, p. 98 ;---*Perhaps,* THE SUCCESSION is lefs effential to WHEAT, than are *Manure* and *Tillage.*

By Obfervation, p. 98 ;---CLOVER, WHEAT—was a moft eligible Succeffion, on a fandy, gravelly Loam.

SOIL-PROCESS.

THIS laborious department of the Vegetable Management prepares the SOIL for the reception of the SEED ; and is the firft operation of the PLOW-MANAGEMENT.

By Experimemt, No. 3 ;---*Mixgrafs* was good in proportion to the nnenefs of the TILTH.

By Experiment, No. 8 ;---A SUMMER FALLOW of fix plowings gave a larger Crop of *Wheat,* and a cleaner Quondal, than a FALLOW-CROP with two hoings, a weeding, and five plowings.

By Experiment, No. 12 ;---One additional CROSS-PLOWING encreafed the Crop confiderably.

By Experiment, No. 15 ;---ACCLIVATING GRAVEL was obvioufly beneficial to *Wheat.*

By Experiment, No. 30 ;---DISCUMBERING the Surface of a *Wheat-Quondal* on a GRAVELLY LOAM, before plowing, rendered it exceedingly *fallowy.*

By Experiment, No 67 ;---A FALLOW which has been broken up in. Autumn, by *trenching,* fhould remain in *open Trenches* until dry weather fet in.

By

By *Obfervation, p.* 48 ;---Perhaps, ROLLING FALLOWS very hard is good management.

By *Obfervation, p.* 48 ;---The Crops and the Quondals were nearly in proportion to the QUANTITY OF TILLAGE.

In *Obfervation, p.* 49 ;----ACCLIVATING POROUS SOILS is fully treated of.

By *Obfervation, p.* 50 ;---Perhaps HALF-ROD RIDGES are the moſt eligible Beds for a *Clay*, or for a *Gravel*.

By *Obfervation, p.* 92 ;---SOD-BURYING a *Clover-Ley* for *Peafe* was eligible.

By *Obfervation, p.* 99 ;---SOD-BURYING a *Clover-Ley* for *Wheat* was good management.

In *Obfervations, p.* 106, 107 ;---TRENCHING FALLOW appears in a favourable light.

SEED-PROCESS.

THIS Diviſion of the PLOW-MANAGEMENT is ſubdiviſible into

Time of ſowing,　　　　　Quantity of Seed,
Preparation of the Seed,　Covering,
Mode of ſowing,　　　　　Adjuſting.

Which Subdiviſions are preſerved in the article WHEAT, 1777. But the general Inferences which I have been able to draw with reſpect to the Seed-proceſs, during the laſt two years practice, will not warrant here a ſeparate treatment of its Subdiviſions.

The TIME OF SOWING, taken in a general ſenſe, is perhaps the moſt important department of this Proceſs: as for inſtance, the ſame elements and operations which would give a good Crop of *Wheat*, provided the Seed was ſown in Autumn, would, if the time

　　　　　　　　　　　　　　　　　　　　of

SEED PROCESS.

of fowing was deferred until late in the Spring, be only productive of *Straw* and *Chaff*.   And it is obfervable, that even a few days
difference in the time of fowing, either in Autumn or in Spring,
fometimes make an obvious difference in the Crop.

To afcertain precifely the propereft time of fowing, is perhaps a
difficult tafk ; and, if afcertained, could not in a general fenfe be
ufeful to a *Farmer* (more efpecially the *ftiff-land Farmer*) ; for his
Beafts of Labour being in due proportion to the fize of his Farm,
he muft in fome meafure be guided by the OTHER OPERATIONS
incident to his Profeffion, as alfo by the STATE OF HIS SOIL, with
refpect to its *Moifture, Tilth,* &c. and cannot, like a *Gardener of a
few rods,* or a *Cultivator of a few loamy acres,* fow on the day and
hour he pleafes.

LINNÆUS neverthelefs merited applaufe, when he recommended
to his Difciples the obfervance of the FOLIATION OF TREES, in
hopes that fome general benefit might thereby arife to mankind ;
and more particularly to the *Farmer,* with refpect to the TIME OF
SOWING.

Mr. STILLINGFLEET, who publifhed tranflations of feveral valuable papers read at the Univerfity of UPSAL *, (among which is
one on the FOLIATION OF TREES) has given a CALENDAR OF FLORA,
which he formed in NORFOLK in the year 1755.   His motives for
publifhing it were two-fold ; firft, to compare it with a SWEDISH
CALENDAR made the fame year ; and next, to induce his countrymen
to make fimilar obfervations.

' *Stillingfleet's Tracts* falling into my hands this Spring, I was ftruck
with the different degrees of forwardnefs of the fpring of 1755,
and that of this year, 1779 ; and was led, without any particular
intention, to mark down this difference in the margin of Mr.
Stillingfleet's Calendar.

* Of which the great Linnæus was prefident many years,

being

SEED-PROCESS.

Finding the taſk agreeable, and the comparifon appearing intereſting, I was induced to make out a Catalogue of ſuch *Vegetables* and
Birds as were moſt likely to fall under my obſervation. But it
being towards the latter end of February before I made out this
liſt, and became accurate in my obſervation, ſome of the early
plants had flipt my notice entirely, and others I was obliged to
regiſter from recollection. However, I did not inſert any but
ſuch as I could recollect with a ſufficient degree of accuracy : and
Mr. Sɹillingſleet having obſerved that the Spring of 1755 was *backward*, and this Spring of 1779 being *uncommonly forward*, I am
induced to offer to the Public the following REGISTER, as being
exhibitory of a ſtriking contraſt between a *late* and an *early*
Spring ; as well as to introduce with greater propriety the inferences
I drew from it.

But prefatory to the Regiſter, it is proper to mention the climatures in which the obſervations were ſeverally made.

Mr. Stillingfleet's were made in lat. 52° 45′. And he deſcribes
the aſpect, ſoil, &c. thus : " All the country about is a dead flat ;
" on one ſide is a barren black heath, on the other a light ſandy
" Loam ; partly tilled, partly paſture land ſheltered with very
" fine Groves."

Mine were made on various Soils, and in different climatures :
prior to the 25 March, they were made in the ſituation minutely
deſcribed in the *Introduction to the Experiments :* ſince that time I
have endeavoured to take the Mean, between the highly-cultivated
country near Town and the *neglected*, though delightful, wilds of
NORWOOD.

A

A Comparative View of the PROGRESS *of* VERNAL VEGETATION *in* ENGLAND, *in the* YEAR 1755, *with that of the* YEAR 1779.

	NORFOLK 1755.	SURRY 1779.
HAZLE Catkins full-blown - -	February 22.	February 10.
Goofeberry began to foliate - -	—— 25.	—— 10.
Sallow Catkins full blown - - -	March 11.	—— 20.
Violets full blown - - - -	—— 28.	—— 14 !
Primrofes full blown - - -	—— 29.	—— 7 !
Birch 'leaves quite out' .. - -	April 1. (*a*)	April 7.
Bramble 'leaves quite out' (*b*) - -	—— 3.	
Cleavers began to fpring - - -	—— 3.	February 21 !
Briar (Dog-Rofe) 'leaves quite out' -	—— 4.	March 4.
Swallow returned - - - -	—— 6.	May 8 ! (*c*)
Filbert 'leaves quite out' - - -	—— 7	—— 25.
Sallow 'leaves quite out' - - -	—— 7. (*d*)	April 7.
Lilac began to leaf - - - -	—— 7.	March 4.
Nightingale began to fing - - -	—— 9. (*e*)	—— 28.

(*a*) *Birch*. This I fhould imagine to be erroneous ; as the obfervation I made this year was very accurate.

(*b*) *Bramble*. It is obfervable, that the Winter being uncommonly mild, many Brambles preferved their foliage and verdure; fo that it was difficult to fay when they came into Spring-leaf.

(*c*) *Swallow*. This was the firft day I faw one ; yet fome ftragglers were *faid* to have been feen a fortnight or three weeks before that time ; but even now (the middle of May) Swallows are ftill rare ! notwithftanding the uncommon forwardnefs of this Spring. May we not fay---*Perhaps,* SWALLOWS *in general* MIGRATE ; but, *perhaps,* fome *few* remain DORMANT in the Ifland during Winter ?

(*d*) *Sallow*. This and the *Filbert* I apprehend are miftakes, not only from this year's obfervation, but from common obfervance of the *Lilac,* which is an earlier fhrub than either of them ; whereas they appear to have foliated, when the *Lilac* is only *beginning* to foliate.

(*e*) *Nightingale*. In 1776 I heard her firft on the 17th April Does not her early finging this year make it probable that fhe either is *not a Bird of Paffage,* or migrates to a lefs diftant country than does the *Swallow ?*

Elm

SEED-PROCESS.

	NORFOLK 1755.	SURRY. 1779.
Elm (narrow-leaved) began to foliate -	*April* 10.	*March* 18.
Turneps in full bloom - - -	—— 15.	—— 25.
Damsons in full bloom - - -	—— 16.	—— 20.
Cuckow began to call - - -	—— 17. (*f*)	*April* 16.
Oak began to leaf - - - -	—— 18. (*g*)	—— 20.
Blackthorn began to blow - - -	—— 18.	*March* 15
Pear-Tree began to blow - - -	—— 18.	—— 18.
Cherry in full blow - - - -	—— 18.	*April* 7.
Hawthorn began to foliate - - -	—— 18.	*March* 15.
Walnut 'leaves quite out' - - -	—— 21. (*h*)	*April* 20.
Wallflower full blown - - -	—— 21.	*March* 28.
Ash ' leaves and flowers quite out ' -	—— 22. (*i*)	*April* 25.
Furze in general bloom - - -	—— 22.	*March* 25.
Lilac in full bloom - - - -	*April* 27.	*April* 7.
Hawthorn began to blow - - -	—— 28.	—— 12.
Cowslips full blown - - - -	*May* 4.	—— 20.
Hawthorn in full blow - - -	—— 10.	*May* 1.

Furze

(*f*) *Cuckow.* In 1776 the Cuckow began to call on the fame day, the 17th April; and in 1779 on the 16th April! Notwithstanding 1755 is said to have been a *backward*, and 1779 is undoubtedly an uncommonly *forward* Spring; yet striking as this is, no inference seems to present itself. In the days of Superstition, indeed, the Cuckow would for half as much have been dubbed a *star-gazer.*

(*g*) *Oak* See the following Observations: this, *perhaps*, was taken from some few *extraordinary early* individual.

(*h*) *Walnut.* This I apprehend is an error, either of the Press or the observer, or both; for this year the leaves of the Walnut, even near London (Hyde-Park), *were not disti* ct, much less ' quite out,' before the 20th. But, *perhaps*, this was likewise taken from some *very forward* individual, and the large (L) has been erroneously joined to it instead of the small (l). See page 178.

(*i*) *Ash.* Here is another instance similar to, or rather more inexplicable than

the

SEED-PROCESS.

An important inference may be drawn from this comparative Register; namely, *the forwardness of Spring in this Country varies upwards of a Month.* In some instances in the former parts of these two Springs there is a disparity of *five or six weeks!* And it is probable, that had it not been for the considerable check which the vegetation of this Spring received in March-April, a similar disparity would have continued until May; and if the approach of Spring is so *casual* as to vary a month or six weeks, we may say with some degree of safety, that the chilling Winds of March-April were merely a *casualty,* which at another time may not happen. And further, perhaps were the Spring of this year compared with the *most backward* Spring which has happened in this Country, a difference of six or eight Weeks might appear.

In the polar Latitudes ; or, to speak of our own Hemisphere, —— in the Northern Climates, where Summers are *short,* vegetation is *rapid* ; in more Southern Latitudes; where Summers are *longer,* vegetation is *slower* in its advances towards maturity. May we infer from this, that *Summers,* whether in the same or in different Climates, *have some certain degree of exertion with respect to vegetation,* which *certain*

the last. Are we to suppose that the Spring of 1779 had received so considerable a check, or that of 1775 so considerable a stimulus in the former part of April as to render them, in the wane of April, of equal forwardness ? The *Lilac* and the *Hawthorn,* and the universal forwardness which now reigns, contradict this supposition. Shall we infer, that *large Trees* are subject to laws of vegetation different from those of *Shrubs* and *Herbs?* The *Elm* flatly contradicts the inference. We may therefore venture to conclude, that there is some error in the registering (and it is not probable that *error* should be so *uniform*), or that the *Oak,* the *Walnut,* and the *Ash,* are guided by laws different from those of the other vegetables which compose the Register; and, indeed, that the *Heat* of the Atmosphere has not any influence on their vernal Vegetation.

degree

degree is fpent during the courfe of *every* Summer ; whether it commences in February, March, or April ; or whether its duration happens to be four, fix, or eight Months ? and that whatever fhare of this exertion is exhaufted on any given part of the Summer, as the former part, for inftance, fo much the remaining parts will be deficient ? This admitted, it follows, that if a Vegetable which requires the *whole Exertion*, be fown after a part of that exertion has been expended, fuch Vegetable did not receive its *proper Time of Sowing*.

If this Hypothefis be well founded, the foregoing inference is very interefting ; for it proves that a Farmer, who confults the *folar Calendar* alone, may miftake his time of fowing a month or fix weeks, whereby he may lofe an effential fhare of that exertion, which, if he were to confult the progrefs of Vegetation, and regulate his Seed-Time by the *terreftrial Calendar*, his Crops would receive.

If this Hypothefis be founded in Error, why fhall we attend to the foliation of Trees ? Why not regulate our TIME OF SOW-ING by an *Aftronomical Ephemeris* ? For the EARTH cannot be fuppofed to have, like the ANIMALS fhe rears,--her *Days of Luft*,---her *Seafons of Love*.

Whether this Hypothefis is or is not well founded, can be decided only by a feries of ACCURATE OBSERVATIONS during a COURSE OF YEARS[*].

In

---

[*] Perhaps, by comparing a minute Regifter of the ftate of the Atmofphere with a minute Regifter of vernal Vegetation, the CAUSE OF VEGETATION might be nearly afcertained. Thus, by a Thermometer (for inftance), calculate how *much* Heat, reckoning from the Winter folftice, is neceffary to bring the Hazle catkin to blow— how *much* to produce Violets, &c. not the *degree* of Heat, at any certain time, but the *quantity* taken collectively, reckoning from fome certain point, as the *freezing-point*. Thus we will fuppofe Violets to require 100 degrees of Heat (namely, degrees above the freezing-point) to bring them to perfection. If the Thermometer were to remain ftationary at 4°. we might expect Violets in 25 days after the

Winter

SEED-PROCESS.

In making thefe Obfervations, however, great Circumfpe&ion is neceffary : not only in the *obferving*, but in the *regiftering* of the appearances; efpecially if the Regifter be intended for *public* infpe&ion. The want of this precaution renders Mr. Stillingfleet's Calendar, fo far as it relates to the *foliation of Trees*, very imperfe&; for he frequently defcribes it by the " leaves quite out" on fome certain day of the month, and regifters this folely by a fingle letter (L), which is too liable to typographical error, to be trufted on a fubje& fo nicely accurate as this ought to be to render it of ordinary ufe.

Befides, nothing can be more indefinite than to fpeak of a Tree's being *quite out*, or in *full Leaf*. As for inftance, the *Buds* of the narrow-leaved ELM this year began to *burft* about the *eighteenth of March*; yet it could not be faid to be in *full Leaf* until the *latter end of April*.

The Elm, it is true, *opened* into Leaf the *feventh of April*; but the *opening of the Leaf* cannot be the period meant by Mr. Stillingfleet, for he has another chara&er for " leaves beginning to open ;" namely, a fmall (l). Nor by " leaves quite out" can Mr. Stillingfleet mean the *full Leaf* here fpoken of; namely, the day

Winter folftice ; if at 2°. in 50 days, &c. Should the Thermometer fall to, or beneath the freezing-point (or perhaps, fome certain point below the freezing-point), we may fuppofe vegetation to be checked, and confequently nothing can be reckoned until the Thermometer rifes again to above that point. The *Weight Moifture, Motion*, &c. of the Atmofphere might be obferved in a fimilar manner. And, by comparing the feveral ftates of the Atmofphere with the progrefs of Vegetation, it might be difcovered, perhaps, that HEAT is not the *only* CAUSE OF VEGETATION.

Thefe refle&ions, tho' in fome meafure fpeculative, arife from the obfervations made on the *Oak, Afh*, and *Walnut*, in notes (g, b, i) page 175. as well as from Mr. BARCK's obfervation in his paper on the Foliation of Trees, with refpe& to a variety of plants which blow in Winter, or early in Spring ; yet " cannot by any care be " brought to flower in Autumn, or after the Summer-folftice." *Stillingfleet's Tra&s*, third *Edition*, page 139.

on

on which the Leaf received its fulleſt width, or largeſt ſize; for he makes the Oak's " leaves quite out" the *ſixth of May*; and it will not even this year be in that ſtate before the *beginning of June*.

Nor is Mr. Stillingfleet's term of " leaves beginning to open" leſs definite; for to re-mention the ELM this year, whoſe BUDS *began to open* the *eighteenth of March*; but they were not *open*: a *majority* of the LEAVES were not *diſtinct* and *ſeparately diſcriminable* until the *ſeventh of April*. It is true, that the cold of March-April retarded the foliation of the Elm this year, but is not every tree *liable* to be checked or retarded in its foliation in every year * ?

If, therefore, the foliation of an *individual* is ſo difficult to communicate, how vague muſt be the idea when we ſpeak of the *Species?* For by accurate obſervation this year, there were ſeveral OAKS in the very ſame ſtate of foliation on the *ſeventh of April*, which the *Majority* of the Oaks of the ſame wood, and the ſame aſpect, were on the *twentieth of April*; and which many others did not reach until the wane of the month. Thus we may ſay, the OAK in 1779 *foliated* the *ſeventh of April*, the *twentieth of April*, or the *thirtieth of April!* And if we ſay that the obſervation was made in latitude 51°. 27'. on a ſtiff, clayey Soil, with a northern aſpect, ſheltered from the eaſt, we ſhall not be leſs unintelligible.

Having thus endeavoured to ſhew the neceſſity of uſing great circumſpection in forming a Regiſter of vernal Vegetation, I will now beg leave to ſubmit ſuch RULES FOR OBSERVING THE FOLIATION OF TREES, as I have drawn from the Experience which I have had this Spring. And firſt, as to *Individuals*.

---

* This is not ſpoken in condemnation of Mr. Stillingfleet's Calendar of Flora; but to ſhew the caution which is requiſite in the obſervance of the Foliation of Trees. Indeed, Mr. Stillingfleet himſelf ſpeaks of the inaccuracies of his Calendar, with a candour and ingenuouſneſs which beſpeak him to have been a man of a uſeful and ſuperior underſtanding; and, *although a man of learning*, his OBSERVATIONS ON GRASSES, as well as his CALENDAR OF FLORA, prove him to have been—— A MAN OF OBSERVATION.

The

The moſt ſtriking ſtage of foliation is, when the BUDS have burſt, or fully expanded themſelves, and the LEAVES, tho' yet *diminutive*, appear *diſtinĚ:* for, prior to this ſtage, the tree, notwithſtanding the *ſwelling* of the Buds, ſtill retains its Winter appearance; but as ſoon as the leaves are diſplayed, the ſpray becomes in a great meaſure obſcured by the verdure of the young leaves, and the ſhrub or tree puts on its vernal aſpeĚ. This proceſs of Nature is begun and finiſhed, in ſome individuals (eſpecially if the Weather be warm and ſhowery) in a few days; and is not only *apparently*, but literally the *foliat on*, or *produĚion of the leaves.* Having marked the *beginning* and the *finiſhing* of the foliation of an individual the *mean*, or *medial Day*, may be taken, and conſidered as the DAY OF FOLIATION of that INDIVIDUAL.

By a ſimilar proceſs, the foliation of the SPECIES might be nearly aſcertained; namely, by marking the *beginning* and the *finiſhing* of the INDIVIDUALS, and by taking the *mean* between the two extremes, and conſidering this as the DAY OF FOLIATION of the SPECIES *.

However, with reſpeĚ to the UTILITY of obſerving the Foliation of Trees, whether we view it in the light of regulating the TIME OF SOWING, or of aſcertaining the FORWARDNESS or BACKWARDNESS of SPRINGS, or of pointing out the CAUSE OF VEGETATION, the obſervations ſhould, I think, be made *every year* on the *ſame* INDIVIDUALS; and whether it be by the *Philoſopher*, or the *Farmer*, he ought to mark out a SUIT OF INDIVIDUALS, and conſider theſe as the *Meters* of VEGETATION, in the ſame manner as he conſiders the BAROMETER, the THERMOMETER, &c. as the *Meters* of the

---

* This might be *checked*, and, if neceſſary, *regulated* by the apparent *majority* which ſhewed itſelf on any particular day; as ſome *excentric* individuals begin to foliate *uncommonly early*, and others *particularly late*; but, by taking the *mean of the two means*, the DAY OF FOLIATION would be aſcertained with a conſiderable degree of accuracy.

SEED-PROCESS.

STATE OF THE ATMOSPHERE*. And let him obferve and mark down the motion, or afcent of the SAP, whether it fhews itfelf in the Bud, the Leaf, the Bloffom, or the Fruit, with the fame exact nefs and care as he does that of the MERCURY; and, by comparing the different columns of his Regifter, obferve what ftate of the Atmofphere *promoted*, and which *retarded* vegetation. And let the Farmer, at Harveft, compare the STATES OF VEGETATION at the different TIMES OF SOWING with the CROPS THEY SEVERALLY PRODUCE, and thus regulate in future the PROPER TIME OF SOWING

It has already been obferved, that a *Farmer by profeffion*, whofe beafts of labour are duly in proportion to the fize of his Farm, cannot attend *nicely* to the foliation of Trees with refpect to his Time of Sowing; but if Springs vary *a month or fix weeks* in their degrees of forwardnefs, it feems reafonable that he fhould not entirely overlook the progrefs of vernal Vegetation Nothing, however, but a feries of obfervation can afcertain the truth of this Hypothefis; for an Hypothefis it muft, as yet, be confidered : and it is probable that the *utmoft* he can *ever* profit by it, will be fomething fimilar to this ; namely,—to *begin* SPRING-SOWING while the *Hazle* (for inftance) *blows*, and finifh while the *Oak foliates*

The *Farmer of a few Acres*, indeed, may perhaps find fomewhat like the following to be a ufeful guide; namely, to fow his PEASE and BEANS while the *Hazle Blows*,---his OATS while the *Sallow blows*,——and his BARLEY while the *Hawthorn blows* †.

Thefe intimations, however, will not I hope be miftaken for PRECEPTS ; but will be confidered only as anticipating what we may, after many years accurate obfervation, expect from the FOLIATION OF TREES, with refpect to the TIME OF SOWING.

* Perhaps, the propereft individual would be thofe of a *middle age* ; as young growing trees, or old decayed ones may *vary* in their Foliations.

† The *Autumnal Seed-time* may perhaps be regulated, in a fimilar manner, by the *falling of Leaves*, the *forming of Catkins*, &c.

I   To

SEED-PROCESS.

To thefe ftrictures on the *Time of Sowing*, I will add the EXPERI-
MENTS and INFERENCES appertaining to the SEED-PROCESS in
general.

Experiments, No. 1, 4;---are on the COVERING of *Clover :* See the
Experiments.

By Experiments, No. 5, 7, 19;---PREPARING THE SEED of *Wheat*
by pickling, *feemed* to be rather injurious, than beneficial to the
Crop.

Experiment, No. 9;---MODE OF SOWING: Two and half bufhels
of Seed fown *over plit*, gave a better Crop, and a cleaner Quondal,
than the fame quantity fown *under-plit*.

Experiment, No. 13;---MODE OF SOWING: The *ftale Plit plowed
clovery*, and afterwards *fluted*, gave a better Crop than did fowing
over the *frefh plit* which had been *paftured*.

By Experiment, No. 16;---THIN SOWING of *Wheat* on *Gravel* was
fortunate, the Summer proving *wet*.

By Experiment, No. 32;---FLUTING FOR PEABEANS gave an evenner
Crop, and a cleaner Quondal, than SOWING OVER THE FRESH PLIT did.

Experiments, No. 35, 36; MODE OF SOWING *Tare Barley:* It was
not material whether the Plits were *broken,* or left *whole,* or *fluted*.

Experiment, No. 55;---TIME OF SOWING: Re-fowing *Clover,* when
Oats were in haw, was of no perceptible fervice.

Experiment, No. 61;---PREPARING THE SEED: No advantage arofe
from *pickling* the Seed of *Wheat*.

Obfervation, page 19;---TIME OF SOWING *Clover :* Perhaps, it is
more certain to fow after the Corn is up, than on the naked furface.

Obfervation, p. 37;---MODE OF SOWING: *Peafe* fhould be buried
deep on a dry foil.

Obfervation, p. 42;---COVERING LEY-GRASS SEEDS: Perhaps it
is good management to make a practice of *hand-raking them in*.

Obfervation, p. 52;---TIME OF SOWING: Perhaps, fow poor land
early,

early, rich land late; *beginning* with that which is poor, and *finishing* with that which is in heart.

Observation, p. 54;---MODE OF SOWING: Perhaps, on cold, wet land, two bushels of Wheat sown *over the fresh Plit,* is equivalent to three bushels *plowed in.*

By Observation, p 55;---The QUANTITY OF SEED depends greatly on the *Mode* and *Time of Sowing.*

Observation, p. 102 ; COVERING :---One shilling saved in harrowing, was Twenty lost in the Crop.

VEGETIZING-PROCESS.

THIS stage of the vegetable management *commences* where the SEED-PROCESS *finishes*; namely, immediately after the seed has been *sown* and *covered,* and the soil *adjusted,* in such a manner, as to be able to disburden itself of superfluous moisture.

The business of this department is to *defend the rising Vegetable,* and *to aid by* ART its NATURAL VEGETATION ; by removing whatever may retard or obstruct, and by *adding* what may promote its advances to maturity. It is *the Process appertaining to the* REARING OF VEGETABLES; and consequently, when they have reached the state of maturity, deemed most beneficial to the Agriculturist. The VEGETIZING PROCESS is *closed* by *Harvesting, Haying, Verdaging,* &c.

This department of Husbandry has perhaps been *too little* attended to by *professional Farmers,* and perhaps *too much* by *Cultivators of a few loamy Acres.*

By Experiment, No. 40; ROLLING PEABEANS, when they were in broad Leaf, was not injurious to the Plants.

By Experiment, No. 42 ;---ROLLING WHEAT ON CLAY was not beneficial to the Crop.

3

By

By Experiment, No. 43 ;---CHECKING RANK WHEAT, in April, *the Summer proving wet,* was beneficial to the Crop; but injurious to the Quondal.

By Experiments, No. 62, 63 ;---ROLLING BARLEY, when fix or eight inches high, was not materially injurious to the Crop, But fee thefe Experiments.

By Experiment, No. 66 ;---HARROWING OATS, when fix or eight inches high, was not detrimental to the Crop.

By Obfervation, p. 23 ;---DISWEEDING MEADOWS, is a valuable operation, performed at a fmall expence.

By Obfervation, p. 57 ;----*Perhaps,* CHECKING WHEAT, which is not *very rank,* is bad management.

VEGETABLE-PROCESS.

THIS ftage of the management of Vegetables receives them as foon as they have acquired a *proper* degree of maturity, and either *immediately applies them to* PASTURAGE, VERDAGE, MELIORATION ; or *lays them up* for FUTURE USE ; and is, in an enlarged fenfe, the HARVEST-PROCESS.

By Experiments, No. 53, 54 ;---MAKING MEADOW HAY in fmall Cocks, the Weather being wet, was eligible.

By Experiment, No. 59 ;---HARVESTING OATS : They fhould not be cocked until the day they be carried.

By Obfervation, p. 39 ;---THE TIME OF CUTTING, whether Grafs or Corn, can only be determined by attentive obfervation.

In the Obfervations, p. 86, 87 ;---MEADOW HAY-MAKING is particularly treated of.

In Obfervations, p. 92, 93 ; -----HARVESTING PEASE is fully treated of.

FARM-

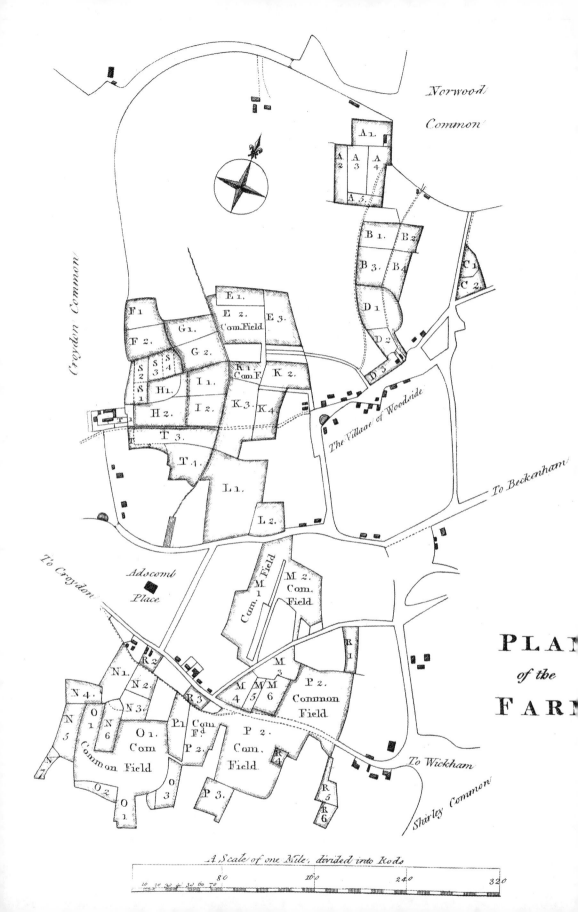

Norwood Common

Croydon Common

A 1.
A 2.
A 3.
A 4.
A 5.

B 1.
B 2.
B 3.
B 4.

C 1.
C 2.

D 1.
D 2.
D 3.

E 1.
E 2. Com. Field.
E 3.

F 1.
F 2.
G 1.
G 2.
S 2.
S 3.
S 4.
S 1.
H 1.
I 1.
I 2.
K 1. Com. F.
K 2.
K 3.
K 4.

H 2.
T 3.
T 4.

L 1.
L 2.

The Village of Woodside

To Beckenham

Adscomb Place

Com. Field

M 1.
M 2. Com. Field.
M 3.
M 4.
M 5.
M 6.
R 1.
R 2.
R 3.
P 2.
Common Field

To Croydon

N 1.
N 2.
N 3.
N 4.
N 5.
N 6.
N 7.
O 1.
O 6.
O 1. Com on Field
O 2.
O 1.
O 3.
P 1.
Com. Fd.
P 2.
P 2. Com. Field.
P 3.
R 4.
R 5.
R 6.

To Wickham

Shirley Common

PLAN
of the
FARM

A Scale of one Mile, divided into Rods

10 20 30 40 50 60 70   80   100   240   320

[185]

FARM-YARD MANAGEMENT.
M A R K E T S.
A C C O U N T S.

THESE three departments of THE VEGETABLE MANAGEMENT appertain to every diſtinct AGRICULTURAL VEGETABLE, as much as do the *Soil*, the *Seed*, and the *Vegetable Proceſſes*; for the VEGETABLE MANAGEMENT is not *cloſed*, until the *produce* has been *conſumed*, or *diſpoſed of*; nor the enquiry reſpecting it *finiſhed*, until the *Profit* or *Loſs*, ariſing from each diſtinct Vegetable, be aſcertained.

The FARM-YARD MANAGEMENT *, when applied to *Vegetables*, includes the BARN and STACK-YARD Managements, ſo far as they relate to *conſumption* or *ſale*: And the *actual Yield* appertains to this Head, as the groſs produce and *eſtimated Yield* do to the VEGETABLE PROCESS: conſequently, by a *comparative view*, RULES for the future ESTIMATION OF YIELD will be drawn.

Under this Head, too, the quantity of FODDER *conſumed*; the number of LIVE STOCK it *ſupported*, and the quantity of DUNG *produced*, &c. &c. ſhould be minutely regiſtered †.

Under

* This confined phraſe is made uſe of to avoid innovation. The proceſs here meant by it, receives the Vegetable from the Harveſt-, or Vegetable-Proceſs, and either prepares it for *Sale*, conſumes it as *Fodder*, or reduces it to *Manure*.

† When an employment becomes *familiar*, its ſeveral operations, more eſpecially its MINUTIÆ, ſlide away imperceptibly, without leaving an impreſſion on the mind; and it would, perhaps, be difficult for a man who has been trained up from his infancy in Farming, to make a Series of Minutes on the FARM-YARD MANAGEMENT.

This reflection ariſes from the total want of freſh information, ſince July 1777,

D d

relative

MARKETS.

Under the article Market *, the *pecuniary Produce* of such parts of each Vegetable as have been *sold*, will of course be ascertained ; and the useful incidents attending the *Sale* should be registered.

Under Accounts, the *gross Amount of Produce* should first be calculated ; by bringing the real amount received for the part *sold*, and the estimated amount of that *consumed*, into one sum. The Gross Amount being thus ascertained, the several charges and expences incurred, whether arising from *rent and taxes* (including *tithes) Manure, Seed,* or *Labour* †, should be brought together ; and thus having found the *Gross Amount*, and the *Gross Charges*, the Profit or Loss arising from the cultivation of each Vegetable, will consequently appear ; from which, Rules for the future Management of Vegetables may with certainty be drawn.

relative to this Head ; as well as to Servants, Implements, &c. &c. And, indeed, the remark may be extended to almost every other department, whether *principal* or *minutial* ; for had I not fallen upon a *systematic* method of Record, the incidents collected by *miscellaneous Minutes* would have been very inconsiderable. This idea is strengthened by the Minutes of Agriculture, which are numerous in the Springs of 1775 and 1776, but few in that of 1777.

How necessary then is an annual systematic Register (in which the business of the Farm becomes mechanically recorded, with an almost mathematical certainty) to *the mere Adept in Agriculture* ; and how peculiarly useful are miscellaneous Minutes to the *novitial Farmer* ? And I will venture to add, that every Man who will, while the business of Farming is a *new*, and consequently a *striking* subject to him, minute and communicate the interesting incidents which present themselves, relative to the minutial Management ; and give his Reasons for adopting or rejecting the particular Operations ; will be rendering an acceptable service to this intricate, yet important department of Agriculture.

* This term, too, must be taken in an enlarged sense ; as it includes every thing *carried off*, and every particular relative to *sale*, whether at *Market* or elsewhere.

† See Accounting, in the Digest of the Minutes of Agriculture, and the *Minutes* there referred to.

The

The Reader who has perufed the prefixed *Advertifement*, will not afk why thefe three articles are not feverally treated of, under each refpective Vegetable, as is here recommended. Indeed, the two latter, and efpecially the laft, is of fo *private* a nature, as few men, *under ordinary circumftances*, would think fit to *publifh*. It need not to be obferved, however, that the whole enquiry, without *minute Accounts*, is in fome degree *conjectural*, while *thefe* point out the proprieties and improprieties of management with MATHE-MATICAL CERTAINTY. And although few Men may think proper to *publifh* their profits and loffes by Farming; yet it is effentially neceffary to common prudence, that every Man fhould *know* them.

Thus I have given a SYSTEMATIC VIEW of the principal VEGE-TABLES which occurred to my practice in the years 1777 and 1778. And I have alfo treated, ABSTRACTLY, of the feveral ELE-MENTS, AGENTS, and PROCESSES appertaining to the VEGETABLE MANAGEMENT.

LIVE-STOCK.

I HAVE two reafons for confining this publication to the VEGE-TABLE-Management alone.

Firft, The experience which I have had with refpect to the ANI-MAL-Management, fince July 1777, is too inconfiderable to be an object of public notice; efpecially as the adjoining *Commons* have affifted the *Farm* in fupporting the LIVE-STOCK; and confequently no decifive information could be drawn. And fecondly, I have never paid that minute attention to the ANIMAL, as I have

to

LIVE-STOCK:

to the VEGETABLE, and more efpecially the PLOW, management; partly from the nature and fituation of the Farm; and partly from a principle of confining, as much as poffible, my whole attention to an object, which I found was alone fufficient to employ it; with a thorough conviction that I fhould thereby fooner acquire a competent knowledge of that particular object. And what confirms me in my filence, here, with refpect to the LEY and the ANIMAL Managements, I flatter myfelf that I fhall very foon have an opportunity of ftudying thefe departments of Hufbandry by an enlarged Plan.

At prefent, however, I will venture to recommend an *annual* (or more frequent) SYSTEMATIC REVIEW of the feveral *Species* of agricultural Animals or LIVE-STOCK; as alfo an ABSTRACT View of the feveral *Proceffes* appertaining thereto; in a fimilar manner to that which I have here taken of the feveral fpecies of VEGETABLE, and their various proceffes. Thus, under the feparate heads of

HORSES,   SWINE,
CATTLE,   POULTRY,
SHEEP,   BEES,

enumerate the INCIDENTS, EXPERIMENTS, and THEORETICAL REFLECTIONS appertaining to the feveral ftages of their management; and afterwards take an ABSTRACT and GENERAL VIEW of the refpective departments; as

BREEDING,   FATTING,
REARING,   MARKETS,
DAIRYING,   ACCOUNTS:

And having thus remarked the proprieties and improprieties of management, and afcertained the profit and lofs arifing from each fpecies, draw RULES for the future MANAGEMENT OF ANIMALS.

RETROSPECT

RETROSPECT

OF THE

FIVE YEARS' MANAGEMENT.

THE TWO YEARS' PRACTICE which are the immediate object of this publication, and which, with the MINUTES OF AGRICULTURE, contain a minute detail of FIVE YEARS' EXPERIENCE, wherein my *fucceffes* and *mifcarriages* have been difplayed with equal *nakednefs*, being now clofed; and as this detail contains not only A HISTORY OF THE ACTUAL INCIDENTS OF A FARM, but alfo THE PROGRESS OF AGRICULTURAL KNOWLEDGE, as it arifes from SELF-EXPERIENCE; I will here take a RETROSPECT OF THE FIVE YEARS' MANAGEMENT, and not only point out what, as a *Student*, I *have* done, but alfo what, *after five Years attentive application*, I now think I *ought* to have done.

In executing this, I will firft give *a general view of the Succeffion*, and then enter into a feparate examination of the management of each Field

In juftice to myfelf, however, I muft previoufly obferve, that *the firft Years' Crops were not put in under my particular direction*; fome of them having been fown before I had any *fhare* in the management; and, being principally in London until after Spring feed-time, the reft were, in a great meafure, left to the difcretion of a Bailiff. The 18th of July 1774, I took upon myfelf the *entire direction*, as is fully explained at the commencement of the MINUTES.

A GE

RETROSPECT.

A GENERAL VIEW of the SUCCESSION.

N. B. The *Figures* denote the Number of Plowings which were given for the Crops to which they are respectively prefixed ; *a*, signifies a *partial* or a *flight* dressing ; *b*, a *middling* quantity ; *c*, a *large* quantity of Manure.

		1774.	1775.	1776.	1777.	1778.
A.	1.	Fallow.	5. a Wheat.	1. Oats.	Mixgrafs.	Mixgrafs.
	2.	Fallow.	Fallow.	8. Oats.	Mixgrafs.	
	3.	4. b Wheat.	1. Oats.	1. Oats.	a Mixgrafs.	
	4.	1. a FallowCrop.	4. Wheat.	Fallow.	5. Oats.	
	5.	4. b Wheat.	1. Oats.	Fallow.	5. Oats.	
B.	1.	1. a Oats.	Mixgrafs.	Mixgrafs.	Mixgrafs.	Fallow.
	2.	1. a Oats.	Mixgrafs.	Mixgrafs.	Mixgrafs.	Fallow.
	3	Fallow.	Fallow.	8. a. Wheat.	1. Oats.	Clover.
	4.	1. a Oats.	1. Beans.	Fallow.	6. Oats.	Clover.
C.	1.	1. Oats.	Fallow.	5. Wheat.	b Mixgrafs.	Mixgrafs.
	2.	1. a Oats.	Mixgrafs.	Mixgrafs.	Mixgrafs.	
D.	1.	1. Oats.	Mixgrafs.	Mixgrafs.	Mixgrafs.	Mixgrafs.
	2.	Meadow.	Meadow.	Meadow.	a Meadow.	Meadow.
	3.	Pasture.	Meadow.	Meadow.	Meadow.	Meadow.
E.	1.	5. Wheat.	4. Tare-Barley.	1. Oats.	given up.	
	2.	1. Oats.	4. Barley.	1. Oats.		
	3.	1. Oats.	4. Barley.	1. Oats.		
F.	1.	b Rye and Turn.	4. Barley.	1. Tarebarley	Fallow	5 b } Barley.
	2.	c Potatoes. (1)	5. Wheat.	1. Tares.	3. Tarebarl. (2)	4 a
G.	1.	1. Wheat.	Fallow.	5. Seed-Tares.	1. Oats.	Fallow.
	2.	Rye-Grafs.	4. Wheat.	1. Peafe, &c.	3. a Oats.	
H.	1.	a Barley.	1. Tares.	4 Wheat.	c Clover.	1 Peafe.
	2.	Barley.	Fallow.	6 Wheat.	c Clover.	
I.	1.	Rye-Grafs.	3. a Wheat.	3. Barley.	b Clover.	1. Wheat.
	2.	Beans.	3. Wheat, &c.	3. b Wheat.	c Clover.	
K.	1.	Meadow.	Meadow.	Meadow.	Meadow.	Meadow.
	2.	3. Barley.	Mixgrafs.	b Mixgrafs.	Pasture.	Mixgrafs.
	3.	Meadow.	b Meadow.	Meadow.	Pasture.	Meadow.
	4.	Rye-Grafs.	1. Beans.	1. Wheat.	Fallow.	6. b Oats.
L.	1.	Rye-Grafs.	1. Oats.	Fall. & F. C.	5. b Wheat.	1. Oats.
	2.	Rye-Grafs.	1. Oats.	1. Beans.	3. b Wheat.	
M.	1.	Various.	a Peafe.	4. a Wheat.	1. Tarebarley.	4.
	2.	Various.	b Tares.	4. Wheat.	1. Tarebarley.	4.
	3.	Rye Grafs & Clo.	c R. Gr. & Clo.	4. Wheat.	1. Peabeans.	4. Barley.
	4.		c Various.	4. Wheat, &c.	1. Peabeans,&c.	4.
	5.	Drilled Peafe.	Fallow.	5. b Wheat.	1. Peabeans.	4.
	6.	Rye Grafs, &c.	c R. Grafs, &c.	4. Wheat.	1. Peabeans.	5.

(1) Planted in the Interfurrows, by way of a *Compost Fallow.* See MINUTES of 17 October, 1774.
(2) In order to clafs it with F 1.

		1774.	1775.	1776.	1777.	1778.
N.	1.	b Rye-Grafs.	Rye Gr. & Fall.	4. Wheat.	given up.	
	2.	1. Peafe.	3. Wheat.	1. Tares.		
	3.	2.Oats.	4. Wheat.	1. Tares.		
	4.	Rye-Grafs&Clo.	1. a Tares.	4. Barley.		
	5.	3. Barley.	1. Tares.	4. Barley.	Clover, &c.	1. b Wheat.
	6.	a Peafe.	3. Wheat.	4. Barley.	Clover.	1. b 1. Wheat.
	7.	a Wheat.	1. Tares.	Mixgrafs.	Mixgrafs.	Mixgrafs.
O.	1.	Various.	4. Barley.	Clover.	a Clover.	1. aWheat, &c.
	2.	a Wheat.	4. Tares.	Oats.	Clover.	Clover.
	3.	a Wheat, &c.	1. Oats.	a Clover, &c.	1. Tares.	given up.
P.	1.	Rye-Grafs.	b Rye-Grafs.	R.Grafs & Fall.	4. Wheat, &c.	1. Peafe, &c.
	2.	Various.	4. Barley.	a Clover.	1.a Wheat.	Fallow.
	3.	3. Barley.	1. Oats.	Fall. & Clo.	b Wheat.	1.Peafe.
R.		Mead. & Pafture	Mead. & Paft.	a Mead. & Paft	Mead. & Paft.	Mead. & Paft.
S.	1.	Turnips.	b Cabbages.	1.Barley.	b Cabbages.	1. Barley.
	2.	Turnips.	Carr. & Pota.	b Potatoes.	1. Wheat.	1. Tares.
	3.	Beans.	3. Wheat.	b Turnips.	b Potatoes.	1. Wheat.
	4.	Beans.	3. Wheat.	b Cabbages.	Turnips.	Rye & Turn.
	5.	Various.	Various.	Various.	Tares.	Rye.
T.	3	Fallow Crop.	5. Barley.	c Pafture.	Mixgrafs.	Mixgrafs.
	4.	Pafture.	Pafture.	Pafture.	Meadow.	Meadow.

DIVISION A.

THIS Divifion lying at a diftance, and the road to it being exceedingly bad in the Spring of the year; very little manure to be had in the neighbourhood; and the fields deftitute of water; it was wholly unfit for *Tillage*, or *pafturage for Cattle*. The only profitable purpofe to which it could be put, was to *ley* it; either as an affiftant *fheep-pafture* to the Common, or to mow for *Hay*. But the Common (principally for want of draining) is very liable to give the *Rot*; efpecially to fheep which are not kept conftantly upon it: it only remained therefore to convert this Divifion into Grafs-land for *Hay*. And this idea was purfued; but it was not executed properly; for the unexpired term of the leafe of this Divifion was too fhort to admit of its being converted into *Meadow*. Inftead therefore of leying any part of it with *fine Graffes*, it ought, throughout, to have been at once reduced to a profitable *temporary ley*, by fowing. *ftrong Graffes*, which the firft year would have afforded a burden of Hay. It ought to have been univerfally fown with *Cow-Grafs*, or with a mixture of *Cow-Grafs and Rye-Grafs*.

Befides,

Befides, the *progrefs of leying* A 1, 2, 3, was injudicious. A 1. ought not to have been *Wheat*, but *Oats*, in 1775; for the Summer of 1774 being very unfavourable to Fallow, this Field was not fufficiently clean to have been *leyed*, even with the Wheat, much lefs with a crop of Oats after it : It ought to have been landed-up in the Autumn of 1774, and to have received a late Spring-plowing for *Oats and Cow-grafs* in the Spring of 1775. A 2. was uncommonly foul, and out of heart; and fallowing it two years for *Oats and Ley graffes* was moft eligible management ; but, inftead of *Hayloft-feeds*, it fhould have been fown with *Cow-grafs and Rye-grafs.* As to A 3. the mifmanagement commenced with not fowing grafs-feeds over the Wheat in the Spring of 1774. To rectify this, it ought to have been *Fallow-Crop* in 1775, for *Oats and Mixgrafs* in 1776. A 4. and 5. were leyed in a proper manner, and their Crops in 1778 are pofitive proof of the eligibility of the management they received (fee page 80). They ought, however, to have been leyed earlier. A 4. like A 1. ought to have been leyed in 1775 ; and A 5. like A 3. ought to have been fown with Grafs-feeds in 1774.

This Divifion having been conftantly plowed for a courfe of years, there would have been no danger of the Ley-graffes going off in lefs than five or fix years ; when, with the affiftance of the little manure which might have been collected in the neighbourhood, they would, probably, *after this reft from the Plow,* have afforded two or three good crops of *Corn* or *Pulfe.*

DIVISIONS B, and C.

What has been faid of the Divifion A, is applicable to thefe Divifions : they ought, as foon as conveniency would have permitted, to have been *thoroughly cleanfed* for Oats and *ftrong* Ley-graffes.

The

The flight mixture of Cow grafs, which was fown in B 1, 2. and C 2. lafted, notwithstanding the foulnefs of the foil, four years. And had thefe Fields been leyed with *Fallow* inftead of *one Plowing*, the management would have been ftill better.

DIVISION D.

D 1. ftands in the fame predicament as C 2, and B 1, 2. And it muft be obferved, that all thefe fields were leyed too *flat*; they ought to have been acclivated in the manner defcribed in page 82.

The high price of Hay, added to their lying too wide to be convenient for pafture, were the inducements for *mowing* D 2, and 3. every year.

DIVISION E.

This Divifion was fown with *fine* Ley-graffes in 1775; but, having no leafe, the rents being too high, the landlords not inclined to lower them, and the Grafs-feeds not promifing for a Crop, they were plowed in for Oats in 1776, and the land given up.

DIVISION F.

The whole Divifion ought to have been *Barley and Clover* in 1775 *.

DIVISION G.

G 1. was Wheat in 1774, which had been fown on a vile Clover-ley. The whole Divifion ought to have been *Fallow* or *Fallow-Crop* in 1775, for *Oats and Clover* in 1776.

* F 1. was in 1775 part of it fown with *Lucern,* part *white Clover*, part with *Trefoil,* and part with *red Clover* by way of an *experimental Hay-Pafture*; but the graffes in general miffing, the *Lucern* efpecially, they were plowed in for Tare-barley in 1776.

E e

RETROSPECT.

Division H.

The Barley of this Division, notwithstanding the Soil was not clean, should have been succeeded by *Clover and Rye-grass* for 1775, which might have been succeeded by *Oats* in 1776; and those by *Summer-fallow* in 1777, for *Oats and Clover* in 1778.

Division I.

Instead of *Barley* and *Wheat* in 1776, this Division ought to have been *Fallow* or *Fallow Crop* for *Barley and Clover* in 1777.

Thus these four Divisions would have fallen into that regular and profitable succession in 1775, which they only were arriving at in 1778.

Division K.

K 2. was leyed with a tolerable good Barley-*Fallow* in 1774; and the only mismanagement appertaining to this Field was, its being leyed much too *flat* (See MINUTE of 5 December, 1774). Had it been raised into *gentle waves*, I am of opinion that its Crops of Grass would have been much superior to those which *it* gave in its flat, *water-shaken* state. It ought to have been pastured in 1775, this Field being intended for a *perennial Ley*.

K 4. was an old Ley which had been intended for a Meadow; but being leyed entirely *flat* and sown with *Rye-grass*, it was quite *worn-out*, and the Crop in 1774 was scarcely worth mowing. Instead of *Beans in Drills*, it ought to have been *Oats* or *Summer Fallow* in 1775.

Division L.

The whole of this Division was in the same, or a worse, state than K 4. was in. And as these Fields are strikingly exemplary of that unpardonable Ley-management which is frequently practised, 1 will give a sketch of the Processes by which I have been informed these Fields were converted into *Meadows* !

They

They are from fituation *level*; and the fubfoil being *retentive*, they are what is properly termed *wet Land*. L 1. is proverbially fuch: yet in order to convert this Field into a *Meadow*, it was plowed with a *Turn-wreft* Plow, *without Ridge or Furrow*,—harrowed as *flat* as a table *,---fown with *Rye-grafs*, and *mown* for at leaft feven years fuc-ceffively! Add to this, it had, in the Autumn and Winter of 1773, been *poached* in this flat wet ftate by heavy cattle, fo that many parts of it in the Spring of 1774 were in a ftate of abfolute *mortar*!

This Field was intended to have been mown in 1774; but there being no appearance of a Crop, fome Cows were turned into it; and the pafturage it gave might perhaps be equivalent to its *Taxes*; it could not be worth more.

The Divifion L, however, being in the Leafe confidered as *Mea-dow*, it was with fome difficulty that permiffion was obtained to break it up; yet after it was granted, there was not lefs difficulty to get the plows into it. For in 1774, it was May-day before a man could have walked acrofs it, and the middle of May before a team could have worked in it. In order to prevent the like circumftance in the enfuing Spring, the *intended Ridges* were *fet-out*, in Autumn, by fetting two plits back-to-back; the plow-furrows of which, affifted by deep *crofs-furrows*, carried off the Winter's rains; and the Spring proving tolerably dry, the teams were enabled to get upon it fo early as March-April, when the whole was plowed and fown with Oats.

As this Divifion had been feven or eight years in *Grafs*, a good Crop of *Oats* was reafonably expected; but the Soil, partly from its lying flat and being frequently more-or-lefs under water, and

---

* As a proof of the impropriety of plowing *wet* land *flat*; it is told as a fact, that in order to get *a part* of the crop of Oats, with which this Field was leyed, out of the Field (the Autumn being wet), the reft were obliged to be ftrewed under the Wheels, to bear the carriages above-ground!

partly

partly from the poaching of the Cattle, was become a sheet of *glue*; which being plowed with a *plain* Plow, *without burying the Sod*, it was difficult to cover the Seed; and the Division, on a par, gave a shabby Crop: notwithstanding, the Soil is *very good*, as was proved by the succeeding Crops; for, *after it had been mellowed by lying in high Ridges*, it gave a good Fallow Crop, a noble Crop of Wheat, and a very good Crop of Oats. Had this Division, as was intended, been fallow-cropped in 1778 for *Oats and Ley-Grasses* in 1779, and *Pasture* in 1780, it must, in a few years, have necessarily become, what its Soil and situation fit it for, *excellent Meadow*.

Division M.

This is a straggling, aukward Division; and its management, on the whole, could not, perhaps, have been much improved.

Division N.

The Crops from this Division were in general good; and no useful lesson presents itself; except that N 5. being very clean, it ought to have been sown with *Clover* in 1774.

N 3. was sown with *Wheat* in 1775, to reduce it to the same Crop as N 2. as also by way of Experiment: the Crops were equally good (these two Fields having been broke up from an old Ley in 1774); but some of the *Oats* vegetated, and, surviving the Winter, became *Weeds* to the Wheat.

Division O.

This is a straggling Division, and its management very defective. Had it been wholly *Fallow* or *Fallow-Crop* in 1775, instead of being principally *Barley*, the management would have been incomparably more eligible.

Division

DIVISION P.

Nearly the fame may be faid of this Divifion. However, as it would not have been convenient to have fallowed both the Divifions in the fame year, *Clover* fhould not have been fown with the *Barley* of P 2. (which, notwithftanding the endlefs labour it received as a *Spring-fallow*, was ftill foul); it ought to have been, with the reft of the Divifion, *Fallow* or *Fallow-Crop* in 1776.

DIVISION R.

This includes the fcattered upland Meadow, which were mown or paftured as water and other conveniencies pointed out.

DIVISION S.

This confifts of the *Farm-yard Garden-Ground*; and the four fquare Subdivifions were kept in this rotation: *Cabbages, Turnips, Potatoes, Wheat.* The fifth Subdivifion, confifting of the corners, which are inconvenient for tillage, was intended for *Lucern*: the Subfoil, however, being unfound, I had given up the idea of Lucern, and had intended in future to have joined it with the other four Subdivifions, and to have lengthened the rotation by introducing *Clover* between the *Wheat* and *Cabbages*, for two reafons; it would have prepared the Soil, which is too light, for Cabbages; and would have been very convenient as Verdage for the oxen.

DIVISION T.

This Divifion comprehends the *home pafturing-Paddocks.* T 3. affords a ftriking leffon with refpect to *Leying.* See page 83.

TO this particular Examination of each Field or Division, I will add the following general Obfervation.

Had I depended more on *Summer-Fallows*, and lefs on *partial Fallows*; and, inftead of laying *Manure* on a *firm Surface* I had begun earlier to lay it on *the rough Plit of one deep plowing*, I am convinced that my General Management would have been confiderably improved.

Yet, imperfect as it may in feveral inftances have been, I reflect with fatisfaction that it has proved, on the whole, fo free from imperfections. And I will add, let any man who has had the management of a foul, untoward, ftraggling Farm of near 300 Acres of *various* Soils, neither more nor lefs than five years, take a minute, impartial Retrofpect, and fay, whether or not he has had *fome bad Crops*;—and whether or not he could, after five years experience on that particular Soil and Situation, have *mended his management*.

I will take the liberty further to add,—If every intelligent Farmer would take a fimilar Retrofpect, and with candour point out the *Errors* he has committed in the courfe of his Management; and at the fame time fay what with his prefent knowledge he now thinks he *ought to have done*, and confequently, had he the fame line to tread a fecond time, he *would do*, and publifh his fentiments unadulterated; the Agriculture of this country would be confiderably promoted: for, I believe, it will readily be granted, that the impreffions we receive from *Error*, whether of ourfelves or of others, are deeper and more lafting than thofe which we receive from *Precepts*, let them be ever fo ftrongly enforced. Who, for inftance, has not read Dr. Lowth's Effay on Grammar? and who has not profited by it? How few have heard of Ward! and fewer ftill have been benefited by his elaborate work. The former holds out a few *Errors* of others; the latter gives a large Volume of *pofitive Rules in rhime*.

r

HAV-

HAVING taken a view of what *has been*, and of what *ought to have been*, the Management from 1773 to 1778, we will next fee what the feveral Divifions *would have been,----muft in courfe have been*, in 1779.

A. Mixgrafs,	K. Meadow and Mixgrafs,
B. Wheat,	L. Oats and Ley-graffes,
C. Mixgrafs,	M. Clover,
D. Mixgrafs and Meadow,	N. O. Fallow or Fallow-Crop,
F. Clover,	P. Spring-Corn and Clover,
G. Spring-Corn,	R. Meadow and Pafture,
H. Wheat,	S. Garden-ground,
I. Fallow, or Fallow-Crop,	T. Meadow and Pafture.

Therefore;---in Autumn 1779, there would not, there *could* not, have been one foul field (A 1. and 3. excepted) throughout the whole Farms; and, had I commenced Farming with my prefent agricultural knowledge, the Farms would have reached the fame, or a fuperior ftate, in the Autumn of 1777, when each Divifion would have been entirely *claffed*,----thoroughly *cleanfed*,----in high *Tilth*,---tolerable *Heart*,---well ftored with *Ley-graffes*,---and have been brought into a convenient *regular Succeffion*.

This being probably the laft time I fhall addrefs the Public on the fubject of *my own management*, I will purfue ftill farther what I *intended to have done*, provided I had retained in my poffeffion the fmall Farm mentioned in the prefixed Advertifement.

This Farm confifts of the Divifions F, G, H, I, S, and T (fee the PLAN). The four firft are *Arable Divifions*, the fifth *Garden-ground*, and the laft *pafturing Paddocks*. Its *Soil, Subfoil*, and *Afpect*, have been already defcribed in the INTRODUCTION TO THE EXPERIMENTS.

It

RETROSPECT.

Its leading characteriftic is a *low, flat, wet Farm:* yet from an incident mentioned in p. 70 of the DIGEST, it is not *wet* from the nature of its *foil,* but from the *retentivenefs* of its *Subfoil;* for by cutting a deep drain, as a common fhore, through the wetteft part of it, the adjoining Fields were changed from wet and rotten, to firm, dry land: for by opening feveral veins of Gravel, which probably run through the whole Farm, the fuperfluous moifture found vent, and the Soil and Subfoil were thereby rendered *ab-forbent.*

From this Incident, and from evident figns of Gravel on the upper fide of the Farm, from whence it is probable the whole is fed, my firft ftep would have been to have funk a ditch of four or five feet deep, from the N. W. corner of F 1. to the S. W. corner of T 2.

I am of opinion that this alone would have converted thefe dirty, *rotten* Fields, into a *found,* Folding-Farm. If not, my next ftep would have been to have funk fimilar ditches (or ftill deeper where the defcent would have permitted) between *each divifion.* If fome fpringy patches ftill had been left, my intentions were to have re-lieved them by *fub drains* opening into the ditches or *main-drains.*

The next object would have been to have reduced the *four arable Divifions* into *four entire Fields;* to have laid the ridges (which now run eaft and weft) north and fouth *; and to have kept thefe four Divifions, from the beginning to the end of the leafe, in the *regular Succeffion* of FALLOW †, SPRING CORN, LEY, WHEAT.

The *Garden Divifion* was intended to have been kept, invariably, in the rotation mentioned in the foregoing Retrofpect of this Divifion.

The two principal *Paddocks* would have been *hayed* and *paftured* alternately.

---

* See MINUTE of 26 July 1775.

† *Entire* or *partial,* as its ftate of foulnefs might have pointed out.

My

RETROSPECT.

My intended *Stock*—were four full-grown OXEN,—five or fix hardy Cows,—three or four breeding Sows,—a large ftock of POULTRY,—and, if I could have drained the land fo that it would have been able to bear the fold,—*perhaps* a fmall Flock of fuckling EWES.

I have entered into this detail from two motives: firft, to convey Hints to thofe who are poffeffed of, as well as to thofe who are about to be poffeffed of, a fimilar Farm; of which there are many in this Ifland: and next, to convey my fentiments to the Public at large with refpect to the *Divifion and General Management of Farms.* And I beg leave fully to explain myfelf on this important Subject.

I do not mean that every Farm in the Kingdom ought to be divided into fix equal parts, four of which to be thrown into arable Divifions, and the remaining two into pafturing Paddocks and Garden Ground: But I will venture to deliver, *as my prefent Opinion,* that were Farms in general laid out by this, or fome *fimilar* mode of Divifion, the neceffaries of life would be greatly encreafed, and the labour of arable Farmers confiderably diminifhed. Nor do I mean to enforce that the Rotation of *Fallow, Spring-Corn, Ley, Wheat* -is applicable to every foil and fituation in the Ifland; but I will not hefitate to fay, *it is my prefent opinion,* that what is termed *arable* Land fhould be fometimes *Corn,* fometimes *Fallow,* and fometimes *Ley*; and that two *Corn-Crops* ought not to fucceed each other.

The REGULAR SUCCESSION which I have adopted, is the refult of my Experience *in the neighbourhood of London,* and on a Farm where a Flock of *Sheep* could not with propriety be kept.

I again repeat, that the Rotation of *Fallow, Spring-Corn, Ley, Wheat,* is folely the refult of my own Experience (indeed my firft two years Management prove that it could not be a pre-conceived

F f   Idea)

Idea). And I am fully convinced that it is equal, if not far superior, to any other Succeffion whatever, on the Soil and Situation which have been the fubjects of my practice *.

SUMMER-FALLOWING FOR OATS, or other *Spring* Corn, may perhaps, at firft fight, appear to be an idea as *prodigal*, as I apprehend it is *new*; but I am fo fully convinced of its propriety, efpecially when applied to *foul*, *ftiff* Land, that I find myfelf ftrongly defirous to *recommend* it to *every ftiff-land Farmer*.

For (with a proper Application of that Manure which is effentially neceffary to every Rotation and Succeffion Mankind have yet been able to difcover) his Land will, by this Mode of Culture, reach a ftate of *Cleannefs*, *Heart*, and *Tilth*, which, perhaps, taken jointly, no other fpecies of Management could raife it to ; *two Years* Fallow for *Wheat*, not excepted.

For a *late Spring Plowing* to *ftiff* Land, which has been thoroughly cleanfed, and laid-up dry during the Winter, gives it a *Porofity* and *Mellownefs* which are effentially ferviceable to *ftiff* Land ; and which, if it be fown with *Wheat in Autumn*, it neceffarily lofes.

For, befides the hazard of its being *plowed wet*, it is unavoidably *faddened* by the *Winter's Rains*, and thereby receives a *gluey* confiftency, *in proportion to the finenefs of its Tilth*, which is the bane of *ftiff* Land ; and which, if it be fown with *Oats on Fallow* in the Spring, and with *Wheat on Clover-Ley* in the Autumn, it with moral certainty efcapes †.

Before

---

* The *Failure of the Clover* does not obftruct this Rotation ; for it may always be fubftituted by a Crop of *Pulfe*. Befides, if part of the Clover-feed be fown with the Corn, part as foon as the Corn is above-ground, and if thefe two fowings *mifs*, a third fprinkling be immediately given, there is very little danger of the Failure of Clover on Land which is in *high Tilth*, and *tolerable Heart*.

† For the various Reafons I have had, during the courfe of my Practice, for adopting this REGULAR SUCCESSION, fee MINUTES of the 13 July, 23 October, and

RETROSPECT.

BEFORE I clofe this publication, I will beg leave briefly to explain myfelf as an *agricultural* Writer.

It is not—it never was, my Intention to lay down MAXIMS OF AGRICULTURE ; much lefs to difplay to the world— *a complete Body of Hufbandry !* nor to *dictate in any manner,* not even to the STUDENT, much lefs to the ADEPT.

My intentions were in my former publication, as they are in the prefent ;

1ft. To communicate to the Public at large, more efpecially to the *Gentleman* who is *anxious* to become a *Farmer,* a GENUINE HISTORY of the *Bufinefs,* the *Emoluments,* and the *Amufements* of PRIVATE AGRICULTURE.

2d. To communicate fuch *Facts* and *Reflections* as have immediately refulted from *my own Practice,* and to convey to the *intended,* as well as to the *novitial* Farmer, fuch HINTS and CAUTIONS as I have found *ufeful to myfelf.*

And, 3dly. To point out to *every Farmer*—a *fcientific*—a *mathematically certain,* Mode of acquiring agricultural Information, from SELF-PRACTICE.

Briefly,— my Intentions were,—they are, Equally to remove the *vulgar* and the *fafhionable,* the *illiterate* and the *learned* ERRORS OF AGRICULTURE ; to point out a Method of placing in their ftead fuch INCIDENTS and EXPERIMENTS as actual Practice may furnifh ; with fuch INFERENCES and REFLECTIONS as Reafon and actual Obfervations may fuggeft ; and to make an Overture towards fyftematizing thefe Incidents, Experiments, Inferences, and Re-

and 5 November, 1775 ; 6 February, 26 March, 26 April, and 8 September, 1776 ; 28 April, 1777 : DIGEST, page 76 ; and the Head *Succeffion,* page 168, of thefe OBSERVATIONS. And, for my Motives to preferring *large, open,* ARABLE DIVISIONS, fee MINUTES of the 31 December, 1774 ; 27 April, 1775 ; 8 January, and 17 Auguft, 1776.

fiections

flections, agreeably to the dictates of Analyfis and Experience; and, confequently, to promote the hitherto varying ART OF HUSBANDRY to the conftant and invariable PRINCIPLES OF SCIENCE.

With refpect to the DIGEST OF THE MINUTES OF AGRICULTURE, it has been in a great meafure mifunderftood. There is not a *pofitive Inference* from the beginning to the end of that Performance. It is intended as an EXPLANATORY INDEX TO THE MINUTES; and what have been miftaken for *Inferences*, are intended merely as *References*. They were thrown into the Form they appear in, folely for the purpofe of *charaderizing* the MINUTES they feverally refer to, in order that the Reader might not be embarraffed in the References, and turn over the Pages in vain.

After the work was printed, it ftruck me, that I had rendered the DIGEST by this means ambiguous; I therefore entered, in the Approach to the Minutes, a *profeffed caution* againft thefe *apparently pofitive inferences*; and as that precaution is applicable to the prefent performance, I will here tranfcribe it·

"If, in the perufal of the foregoing fheets, the INEXPERIENCED AGRICULTURIST fhould meet with an Implement, a Procefs, or a Plan of Management which he is pleafed with, let him not *hurry* to the Field of Practice. Let him firft confider *deliberately* whether or not it can be affimilated with his own Plan of Management; let him confult *deliberately* his Soil, his Situation, and his Servants; and thus, by a judicious adoption, and an attentive application, fave the credit of himfelf and the PRACTICE OF AGRICULTURE: for the Author wifhes not to lay down *pofitive Rules* for others; and he here enters a proteft againft fuch *didadic* paffages, whether they occur in the MINUTES, the DIGEST, the EXPERIMENTS, or the OBSERVATIONS, as may have efcaped a *Perhaps*; for his meaning is not to enforce *Precepts*, but to convey HINTS."

With

RETROSPECT.

With refpect to the foregoing OBSERVATIONS,-----they are in
tended as a *Model*,----rather than as a *Work*. However, as *Obfervations of Agriculture*, immediately arifing from *actual practice*, they
have afforded numerous *Inferences*; and have fuggefted a variety of
Refolutions of FUTURE MANAGEMENT.

With refpect to the INFERENCES,-----I have divided them into
pofitive and *probable*. But I have already obferved, that even the
former muft not be received as *infallible rules*. See a Note, page
18 and 19.

With refpect to the RESOLUTIONS,---It has already been intimated,
that they are not to be confidered as *pofitive Inferences*; but as principles of future management, refulting from a general knowledge
of the refpective fubjects to which they feverally appertain.

HAVING thus taken an *hiftorical View* of the firft *three Years'*
practice; and having elucidated and amplified the mifcellaneous
Regifter arifing therefrom, by a *copious explanatory Index*:—Having taken a *fyftematic View* of the laft *two Years'* practice; with a
comprehenfive Retrofpect of the *five Years'* management:—And having
endeavoured to explain myfelf, firft as a *practical Farmer*, and
afterwards as an *agricultural* Writer:—I will clofe the publication of my PRACTICE OF AGRICULTURE with an earneft wifh that
it may prove of fervice to Mankind, more efpecially to the
ENGLISH AGRICULTURIST.

THE END.

ALPHABETICAL INDEX.

A.

ACCOUNTS - Page 185
Animal Economy - 114
Arrangement - - 12, 71

B.

Barley - - 26, 105
Beasts of Labour - 165

C.

Cabbages - *Exp.* 51, 52
Clover - Page 16, 75

D.

Division of Farms - 167

F.

Farms - - - 109
Farm-yard Management 185
Fences - - - 167
Foliation of Trees - 174

G.

Gen. View of Sci. Agriculture 1

H.

Harvest-Procefs - - 184

I.

Implements - - 166

L.

Live Stock - - 187

M.

Manure - - 111
Markets - - - 185

Meadow - Page 22, 85
Mixgrafs - - 39, 80

O.

Oats - - - 65, 101
Oxen - - - 165

P.

Peabeans - - 35
Peafe - - - 89
Potatoes - - *Exp.* 47
Produce - Page 70, 108
Prognostical Arrangement 135

R.

Register of Atmosphere 152
Retrospect - - 189

S.

Seed-Procefs - - 171
Servants - - - 165
Soils - - - 110
Soil Procefs - - 170
Succeffion - - - 168

T.

Tare-barley - - 30
Taxes - - - 63

V.

Vegetable Economy - 114
Vegetable Procefs - 184
Vegetizing Procefs - 183

W

Weather - - 114
Wheat - - 45, 95

Printed in the United States
By Bookmasters